International Perspectives on Mathematics Teacher Education

A Volume in:
Research in Mathematics Education

Series Editors:
Denisse R. Thompson
Mary Ann Huntley
Christine Suurtamm

Research in Mathematics Education

Books in This Series:

International Perspectives on Mathematics Teacher Education

Denisse R. Thompson
Christine Suurtamm
Mary Ann Huntley

≡IAP

INFORMATION AGE PUBLISHING, INC.
Waxhaw, NC • www.infoagepub.com

Library of Congress Cataloging-In-Publication Data

The CIP data for this book can be found on the Library of Congress website (loc.gov).

Paperback: 978-1-64802-629-4
Hardcover: 978-1-64802-630-0
E-Book: 978-1-64802-631-7

CONTENTS

PREFACE

Education is an important and essential endeavor for all societies. The large-scale testing of students' achievement undertaken in studies such as the Trends in International Mathematics and Science Study (TIMSS) and Programme for International Student Assessment (PISA) are evidence of countries' interest in considering how the achievement of their students compares to that of other countries, particularly in critical areas such as general literacy, reading, mathematics, and science. It is natural for researchers in participating countries to then delve into the results to investigate potential reasons for differences in achievement, such as differences in curriculum or differences in teacher knowledge. In addition, it is helpful to explore how various countries prepare primary and lower-secondary teachers to teach mathematics.

Differences in the preparation of teachers for all subjects, including mathematics, are typically a reflection of the values of a given society. Within the global mathematics education community, there are many shared values about how teachers can best engage with students in the learning of mathematics, such as the development of conceptual knowledge and problem solving, and attempting to meet the needs of all students. Discussions of shared values and differences have been facilitated by the International Congresses on Mathematical Education (ICME), which have brought educators from all over the world together every four years since the 1960s to share research about the teaching and learning of

International Perspectives on Mathematics Teacher Education, pages vii–xi.
Copyright © 2021 by Information Age Publishing
All rights of reproduction in any form reserved.

mathematics, including of particular content areas (e.g., algebra, geometry, statistics) and the preparation of teachers at different levels (e.g., primary, lower secondary, upper secondary, tertiary).

In addition to quadrennial conferences related to mathematics education, specialized conferences have also focused on the preparation of mathematics teachers. One such conference in 2005, sponsored by the International Commission on Mathematics Instruction (ICMI), focused on the professional education and development of mathematics teachers. One conference paper based on reports from 20 countries participating in the conference states, in part: "We know little about the organization of the opportunities to learn mathematics and mathematics pedagogy offered to prospective and practicing teachers across the world and their relative effectiveness" (Tatto et al., 2009, p. 313). Since then, the Teacher Education and Development Study in Mathematics (TEDS-M; Tatto et al., 2012), supported by the International Association for the Evaluation of Educational Achievement (IEA), which also coordinates TIMSS, was implemented in 2008 to begin to answer such questions. The results of TEDS-M suggest that there is great variation in terms of how mathematics teachers are prepared around the world (Tatto et al., 2012).

To explore this variation in more depth, we solicited chapters from colleagues around the world for this volume on international perspectives regarding mathematics teacher education. This volume complements and expands on information provided by Novotná et al. (2020), who outline several facets of mathematics teacher education in programs around the world (i.e., structure and characteristics, institutions, study programs, teachers of future teachers, characteristics of entrants to mathematics teacher education programs, and professional teachers' competencies).

GENESIS OF THIS BOOK

We conceptualized this volume to be an extension and companion to a previous volume we edited, *International Perspectives on Mathematics Curriculum* (Thompson et al., 2018). In that volume, authors from eight countries shared insights into the intended mathematics curriculum within their respective countries (the Netherlands, France, Finland, Canada, the United States, Brazil, South Africa, and Korea). We found many similarities in curricular expectations across the countries, but also some interesting differences that might be useful to curriculum developers. In the same way, we hope the current volume will lead to useful learning opportunities for teacher educators across the globe by viewing different ways that mathematics teacher education is structured, developed, understood, and implemented in various countries.

For the current volume, we invited mathematics teacher education scholars and researchers in Brazil, Canada, France, Malawi, New Zealand, Singapore, Sweden, and the United States to contribute a chapter to the volume; this collection of scholars represents countries from five continents (Africa, Asia, Europe, North

America, South America) as well as Oceania. We asked authors to provide an overview of mathematics teacher education in their country, to highlight issues and challenges faced in preparing mathematics teachers, and to present how they are addressing these challenges. Although we gave them much latitude in developing their chapters, we were particularly interested in the mathematics content teachers need to understand, the ways that pedagogical approaches are developed, the messages about the nature of mathematics teaching and learning that are seen in preparation programs, and the interfaces between tertiary preparation and school context. More specifically, as a starting point, we provided the following guiding questions to the authors, which were neither meant to be exhausting nor limiting.

- How is initial mathematics teacher preparation conducted in your country? What is the structure of teacher preparation (e.g., number of years, prerequisites)?
- Who is responsible for initial teacher preparation (e.g., tertiary institutions, school districts)?
- What courses are prospective teachers required to take to prepare them to teach mathematics at various levels (e.g., mathematics content, content-specific pedagogy)?
- What clinical experiences are part of initial teacher preparation (e.g., amount of time, connections between university and schools, support for teacher candidates and supervising teachers)?
- What vision of mathematics teaching and learning is portrayed in teacher preparation programs?
- In what ways does research inform your program and/or drive changes in your program?

The chapters developed as a result of our collaboration with the authors have provided perspectives on mathematics teacher education that are intended to broaden the global mathematics education community's understandings of mathematics teacher education. We hope that the detailed information provided by the chapter authors will be useful to researchers around the world engaged in preparing the next generation of mathematics educators, and that readers will agree that the chapter authors have given us much to think about relative to these issues.

STRUCTURE OF THE VOLUME

The volume consists of ten chapters. Chapter 1 provides an orientation to the volume by discussing some of the broad issues that are addressed in all teacher education programs, as well as some of the questions we suggested authors might respond to when discussing mathematics teacher education in their country. These questions may also serve to guide readers' reflections while reading the various chapters.

Chapters 2–9 are the heart of the volume. In each chapter, the authors describe the preparation of mathematics teachers in one specific country (Sweden, France, Malawi, Singapore, New Zealand, Brazil, the United States, and Canada, respectively), telling the preparation story from their perspectives, raising issues and challenges, and, often, how those challenges are faced. Each team of chapter authors includes active researchers or teacher educators in their respective countries.

Chapter 10 provides a look back at the volume, with reflections based on our reading of the chapters and our own perspectives on the guiding questions from Chapter 1, including some facets that we found particularly interesting or unique when reading the chapters. We end the chapter with some considerations to be addressed in the future regarding the preparation of mathematics teachers. The volume concludes with brief biographies of the editors and authors.

UNPRECEDENTED CIRCUMSTANCES

We conceptualized this book on mathematics teacher education from an international perspective in late 2018 and the early part of 2019. The authors wrote and revised their chapters during 2019 and the very early months of 2020. Thus, neither we nor they could have foreseen the events of 2020, specifically the CO-VID-19 pandemic and the social movements related to societal inequities. Each of these major events have impacted how we educate children as well as their future teachers. The chapters are a reflection of teacher education up to and at the time of writing, before these events upended the nature of education in much of the world. In Chapter 10, we will revisit the unprecedented events of 2020 and their potential implications for the future preparation of mathematics teachers.

POTENTIAL AUDIENCES FOR THE BOOK

This volume has the potential to be applicable to mathematics teacher educators and researchers, including those interested in comparative issues in mathematics education. We hope the volume will stimulate conversation about teacher preparation among the wider mathematics education community and provide a means to consider similarities and differences in preparation, and how those similarities and differences may relate to cultural and societal variations across countries. The volume might be used by policy makers interested in comparing mathematics teacher preparation programs, and they may seek to understand how differences might help explain achievement differences that often appear on international assessments. The volume might also be used by professors as a text in a graduate course on preparing future mathematics teacher educators.

ACKNOWLEDGEMENTS

We extend our thanks to the many authors for their hard work in writing chapters for the volume. We appreciate all their efforts in generating drafts and responding to our questions and edits.

In addition, we extend our thanks to George Johnson, President of Information Age Publishing, for his willingness to publish the volume. As co-editors of the series, *Research in Mathematics Education*, we appreciate the freedom he provides us to cultivate volumes related to our professional interests.

REFERENCES

Novotná, J., Moraová, M., & Tatto, M. T. (2020). Mathematics teacher education organization, curriculum, and outcomes. In S. Lerman (Ed.), *Encyclopedia of mathematics education* (pp. 587–593). Springer.

Tatto, M. T., Lerman, S., & Novotná, J. (2009). Overview of teacher education systems across the world. In R. Even & D. Ball (Eds.), *The professional education and development of teachers of mathematics, The 15th ICMI study, New ICMI Study Series,* (Vol. 11, pp. 15–24). Springer.

Tatto, M. T., Peck, R., Schwille, J., Bankov, K., Senk, S. L., Rodriguez, M., Ingvarson, L., Reckase, M., & Rowley, G. (2012). *Policy, practice, and readiness to teach primary and secondary mathematics in 17 countries: Findings from the IEA Teacher Education and Development Study in Mathematics (TEDS-M).* https://www.iea.nl/publications/study-reports/international-reports-iea-studies/policy-practice-and-readiness-teach

Thompson, D. R., Huntley, M. A., & Suurtamm, C. (2018). (Eds.). *International perspectives on mathematics curriculum*. Information Age Publishing.

CHAPTER 1

QUESTIONS FACING MATHEMATICS TEACHER EDUCATION

An Introduction to the Volume

Christine Suurtamm
University of Ottawa

Mary Ann Huntley
Cornell University

Denisse R. Thompson
University of South Florida

In this chapter we situate this volume within the international arena of similar research on mathematics teacher education. We also provide some background information regarding the questions that we provided to authors to prompt their thinking while writing the chapters. These questions might also be useful to consider when reading the chapters in this volume.

This volume highlights international perspectives on the education of mathematics teachers, focusing primarily on preservice preparation. Research suggests that

International Perspectives on Mathematics Teacher Education, pages 1–8.
Copyright © 2021 by Information Age Publishing

teachers have an enormous influence on how mathematics is taught, and thus, how students learn (Krainer et al., 2015; Lerman, 2001). Mathematics is seen by many as an essential competency, and internationally, there is a great deal of effort to provide mathematics education that is accessible to all (Adler et al., 2005). This increasing demand for mathematics competency calls for a similar increasing demand for mathematics teacher competency (Adler et al., 2005). International comparisons of student achievement in mathematics, such as the Trends in Mathematics and Science Study (TIMSS) and Programme for International Student Assessment (PISA), prompt comparisons of the different ways that students are taught and the ways that countries prepare mathematics teachers to teach.

The recent international Teacher Education and Development Study in Mathematics (TEDS-M) suggests there is great variety in terms of how mathematics teachers are prepared (Tatto et al., 2012). TEDS-M focused on differences in the preparation of mathematics teachers, including the structure of programs, the locus of control of preparation programs, the ways in which programs reflect the dominant socio-political situation, and the degree of focus on mathematics. The study included discussion on different policies, practices, and outcomes of mathematics teacher education within each of the participating countries (Krainer et al., 2015). Countries participating in TEDS-M included Botswana, Canada (selected provinces), Chile, Chinese Taipei, Georgia, Germany, Malaysia, Norway, Oman, the Philippines, Poland, the Russian Federation, Singapore, Spain, Switzerland, Thailand, and the United States of America.

This volume includes contributions by authors from Brazil, Canada, France, Malawi, New Zealand, Singapore, Sweden, and the United States, reflecting some of the same countries as TEDS-M as well as other countries, and provides in-depth insider perspectives on mathematics teacher education in these countries. Although some of the same topics as TEDS-M are covered in the chapters of this volume, such as the context, structure, and content of programs, the chapter authors also discuss various issues in mathematics teacher education and the ways programs in each country deal with those issues. We hope these insights into the preparation of mathematics teachers will broaden the mathematics education community's understandings of mathematics teacher education and provide an opportunity for learning from others about how to address common issues and concerns across countries.

In this volume we consider mathematics teacher education broadly, taking into account the context within each country, the mathematics content that teachers are expected to understand, the ways that pedagogical approaches are developed, the messages about the nature of mathematics teaching and learning, the interface between tertiary preparation and school contexts, as well as the challenges, affordances, and constraints that are present in each context. These eight countries are a small sample of the international arena, and thus, this is not a comprehensive study of mathematics teacher education. Furthermore, each chapter represents the perspective of the particular authors of that chapter, and a different set of authors might focus on or highlight different issues facing teacher preparation in that country. Nevertheless, each chapter is authored by those who are well connected with the country they speak of and, taken together, these chapters present a variety of perspectives, issues, and programs that help us think more deeply about how we prepare and support mathematics teachers. Each chapter provides an overview of mathematics teacher preparation, but also highlights challenges that are faced in each country from the authors' perspectives. Each set of chapter authors, to different extents, discusses aspects of their programs that help to address some of the common issues that mathematics teacher educators face internationally, such as recruitment, field experiences, and addressing learners' needs. They also discuss unique mathematics teacher education issues in their country.

Although we did not ask authors to address specific questions, we did provide guiding questions to prompt a variety of ideas to emerge as authors wrote their chapters. Authors did not need to address all questions, and thus, each chapter is unique in its content, perspective, presentation, and focus. The reader may be surprised at the similarities in the issues that are faced as they read the chapters that follow. However, there are differences, as well. Our purpose, though, is not to compare countries in terms of mathematics teacher preparation but rather, to develop a better understanding of how countries deal with the complexity of mathematics teacher preparation and to learn from each other.

In the Preface to this volume, we listed the initial questions provided to authors as they developed an outline of their proposed chapter. As we read the outlines, we expanded on the questions when we provided feedback to the authors. In the following sections, we group these guiding questions into broad categories and provide a brief rationale for the questions. We expect these questions may also help guide your reading of the subsequent chapters.

WHAT IS THE CONTEXT OF SCHOOLING?

Teaching, and thus mathematics teacher education, may look different in different contexts as curricula, resources, populations, class sizes, geographical context, and socio-economic differences influence what quality teaching might look like (Adler et al., 2005). A description of the schooling divisions and associated grade levels and ages, as well as descriptions of the learners, helps to familiarize the reader with the context. Understanding the environment of schooling helps one to understand the teacher education program, which must respect the nature of the learners and the teachers (Borko, 2004). For instance, research suggests that many mathematics classrooms include a range of learners who bring a diversity of linguistic and cultural practices as well as a variety of mathematical competencies (Adler et al., 2005). The chapters may help us to develop an understanding of how different countries prepare teachers to address the needs of students in a range of settings, from culturally homogenous to diverse socio-economic settings. We proposed that the authors might want to consider:

- What is the context of schooling (what grades/ages are considered primary, secondary, etc.; which grades/ages are compulsory), and how does that context play out in the preparation of mathematics teachers?
- Who are the learners and how do programs prepare teachers for addressing the needs of all learners and the different contexts of teaching? In what ways are teacher candidates prepared to address the range of diversity in the classroom and respect the needs of all students?
- What are some ways that programs prepare mathematics teachers for changing school environments (e.g., varying student demographics including second-language learners, technological changes throughout schooling including online access to teaching materials)?

WHO ARE THE TEACHERS?

The availability of teachers may also influence the preparation of teachers. For instance, the extent to which a country produces and retains the mathematics teachers it needs may influence the program structure and requirements. In some cases, where teachers are scarce, alternative programs may need to be put in place. In reading the chapters, attention can be paid to understanding how different countries attend to the challenge of recruitment, preparation, and induction of mathematics teachers in that country. We presented the following questions to the authors:

- To what extent do countries produce the mathematics teachers they need? That is, are there any issues related to shortages or surpluses in availability of mathematics teachers?
- To what extent is mathematics teacher education consistent and/or transferrable across jurisdictions?
- What are ways that mathematics teachers are recruited, taught, and retained to provide a robust mathematics teaching force?
- What are some of the challenges faced in recruitment, preparation, and induction of mathematics teachers and how are these challenges met?

WHO IS INVOLVED IN
MATHEMATICS TEACHER PREPARATION?

TEDS-M highlighted that there are differences in who is responsible for teacher education across countries, ranging from individual school districts, to government-run programs, to accreditation institutions, to universities (Tatto et al., 2012). TEDS-M suggested that some programs occur as part of a university degree whereas others occur post-degree. This variation in teacher education is quite often due to divergent views of the purposes of teacher education and how best to conduct the preparation of teachers (Tatto et al., 2012).

Thus, as you read the chapters, consider who is responsible for, and who determines the program. This might include the ways in which a program addresses the multiple stakeholders, and builds partnerships with stakeholders including relationships between mathematics educators and mathematicians, schools and universities, or policymakers and researchers. In some cases, programs may be heavily tied to teacher certification, with programs focusing primarily on meeting the requirements for certification or preparing teachers to pass high-stakes certification exams. In other cases, academics may have a greater say in designing programs to not only satisfy certification requirements but also satisfy expectations for preparation that are suggested by research. These differences may affect the emphasis of the program. For instance, there could be different views of the theoretical aspects of learning versus experiential aspects. University-school partnerships may also influence mathematics teacher education as they may provide support to teacher candidates and help to provide alignment between the theoretical and practical aspects of mathematics teacher education. Perhaps a program has a reciprocal relationship between schools and teacher education; if so, mathematics teacher education programs support schools at the same time that schools support the program. We suggested that authors of the chapters consider the following:

- Who is responsible for initial teacher preparation (e.g., tertiary institutions, school districts)?
- How are programs for mathematics teacher preparation accredited at state/province/national levels?
- To what extent are there university-school partnerships that influence mathematics teacher education?
- How, if at all, do mathematicians and mathematics teacher educators work together in preparing teachers?

WHAT ARE THE HISTORICAL, SOCIAL, AND POLITICAL INFLUENCES ON TEACHER EDUCATION?

In reading the chapters, we encourage you to consider the historical, socio-economic, and political environments that might be presented and that help to situate the current preparation of mathematics teachers. Political influences and events can also account for shifts in mathematics teacher preparation. These could include national or international student assessments, teacher assessments, or accreditation standards, as well as different political perspectives on the nature and purpose of teaching and teacher preparation. Guiding questions to the authors included:

- What is the historical perspective within which the current preparation of mathematics teachers exists and is enacted?
- In what ways does the political environment impact mathematics teacher preparation and induction?

- What influences or accounts for shifts in mathematics teacher preparation (e.g., national or international student assessments, or assessments for teachers such as TEDS-M if a country participated)?

WHAT IS THE STRUCTURE OF MATHEMATICS TEACHER EDUCATION PROGRAMS?

Research suggests that mathematics teacher education programs take on a variety of different structures (e.g., Tatto et al., 2012). Programs may vary in terms of the number of years, the prerequisite courses, the admission requirements, whether the program is concurrent with an initial degree program or consecutive (following an initial degree), and the time spent in and nature of the practical component of teaching in schools. There may also be different prerequisites to be admitted to the program. These prerequisites and program courses may also differ depending on the division that teacher candidates are being prepared to teach (e.g., elementary versus secondary) and the extent to which research is used to inform these decisions.

As part of the program structure, consideration should be given to who teaches prospective teachers and the ways in which various types of educators might work together. This could include mathematics education researchers, practitioners, as well as mathematicians. We suggested a variety of questions with respect to the structure of programs that authors might be interested in addressing:

- What is the structure of initial mathematics teacher preparation (e.g., number of years, prerequisites, admissions requirements)?
- What alternatives to initial tertiary preparation exist to prepare mathematics teachers for classrooms? Is preparation different for career changers?
- Who teaches within the program?

WHAT IS THE CONTENT OF THE PROGRAMS?

As one reads through the chapters, there are many things to observe in terms of program content. Each country has their own perspective and program to prescribe the courses that prospective teachers are required to take to prepare them to teach mathematics at various levels. Program content focuses on the knowledge that is deemed most important to teach. However, looking across different countries you may see differences in terms of what knowledge is important as well as in the balance of specific mathematical knowledge, pedagogy, knowledge of students and how students learn, educational policy, and components of teaching practice, to name a few (Adler et al., 2005; Tatto et al., 2012). Research also plays a role in designing course content, both in determining what prospective teachers need to learn, and in engaging them in research to inform their practice.

Different jurisdictions place different emphases on and approaches to the development of prospective teachers' mathematics content knowledge. A great deal of research has focused on the mathematics that teachers need to know in order to teach mathematics and suggests that there is a wide range of mathematical understanding among mathematics teacher candidates, both within countries and across countries (Tatto et al., 2012). Teacher candidates have diverse mathematical histories with some having limited experiences and others with deep mathematics learning (Adler et al., 2005). It is interesting to see how programs in different countries attend to mathematics knowledge for teaching; that is, how they develop teachers' understanding of mathematics in ways that help teachers see and name mathematical ideas, as well as strengthening students' mathematical competencies and building students' sense of themselves as mathematics learners (Ball, 2017). Our questions to authors included:

- What courses are prospective teachers required to take to prepare them to teach mathematics at various levels (e.g., mathematics content, content-specific pedagogy), and who teaches these courses (mathematics department faculty or teacher education faculty)?
- What clinical experiences are part of initial teacher preparation (e.g., amount of time, connections between universities and schools, support provided to teacher candidates and supervising teachers)?
- What vision of mathematics teaching and learning is portrayed in teacher preparation programs?
- What role does mathematics play in programs that prepare mathematics teachers?
- What is the role of research in the program? Does research inform the program content? Do prospective teachers engage in research or learn about research on teaching and learning?

WHAT ISSUES DO PROGRAMS ADDRESS AND HOW ARE THEY ADDRESSED?

There are many ways that programs work to prepare mathematics teachers for changing school environments, such as varying student demographics, technological changes, reduced resources, inequities, large class sizes, and remote learning. We encouraged authors to discuss issues their programs face and ways they are addressing them in their programs. We thus asked authors to consider:

- What challenges are faced in preparing mathematics teachers and how do programs address them?
- What are some aspects of the mathematics teacher education program that are unique and/or interesting?
- What are some aspects of the program that are innovative?

SUMMARY

Although we provided this set of guiding questions to authors as they developed their chapters and shared them here to help guide readers, there may be other themes that emerge while reading the chapters that do not fit into these categories or that cut across categories. Readers will find a variety of perspectives and ideas in these chapters that will help all of us think deeply about the purpose and goals of mathematics teacher education and the ways in which mathematics teacher educators and programs realize those goals. In Chapter 10, we will share and summarize some of the issues within the various chapters that we found particularly interesting and informative.

Work on this volume began in late 2018 and early 2019, before the global pandemic of COVID-19 and the unrest that many countries have experienced around racial inequities. We will briefly revisit these issues in Chapter 10. As you read the chapters, you might consider how different programs prepare mathematics teachers to handle teaching in an uncertain world.

REFERENCES

Adler, J., Ball, D., Krainer, K., Lin, F-L., & Novotna, J. (2005). Reflections on an emerging field: Researching mathematics teacher education. *Educational Studies in Mathematics, 60,* 359–381. https://doi.org/10.1007/s10649-005-5072-6

Ball, D. L. (2017). Uncovering the special mathematical work of teaching. In G. Kaiser (Ed.), *Proceedings of the 13th International Congress on Mathematical Education* (pp. 11–34). ICME-13 Monographs. Springer, Cham.

Borko, H. (2004). Professional development and teacher learning: Mapping the terrain. *Educational Researcher, 33*(8), 3–15.

Krainer, K., Hsieh, F-J., Peck, R., & Tatto, M. T. (2015). The TEDS-M: Important issues, results and questions. In S. J. Cho (Ed.), *The Proceedings of the 12th International Congress on Mathematical Education.* Springer, Cham.

Lerman, S. (2001). A review of research perspectives on mathematics teacher education. In F.-L. Lin & T. J. Cooney (Eds.), *Making sense of mathematics teacher education* (pp. 33–52). Springer.

Tatto, M. T., Peck, R., Schwille, J., Bankov, K., Senk, S. L., Rodriquez, M., Ingvarson, L., Reckase, M., & Rowley, G. (2012). *Policy, practice, and readiness to teach primary and secondary mathematics in 17 countries: Findings from the IEA teacher education and development study in mathematics (TEDS-M).* International Association for the Evaluation of Educational Achievement (IEA).

CHAPTER 2

THE CROSSCURRENTS OF SWEDISH MATHEMATICS TEACHER EDUCATION

Iben Maj Christiansen[a], Anette de Ron[a], Andreas Ebbelind[b],
Susanne Engström[c], Susanne Frisk[d], Cecilia Kilhamn[d],
Veronica Jatko Kraft[a], Yvonne Liljekvist[e], Mathias Norqvist[f],
Rimma Nyman[d], Lisa Österling[a], Hanna Palmér[b], Anna Pansell[a],
Astrid Pettersson[a], Kerstin Pettersson[a], Inger Ridderlind[a],
Christina Skodras[d], Kicki Skog[a], and Lovisa Sumpter[a]

As with any programs in teacher education, Swedish mathematics teacher education is influenced by changing political winds, developments in Information and Communication Technology (ICT), culture, history, PISA results, research-based program designs, and a fair amount of passion. Content and outcomes are nationally determined and include the requirement of a strong research foundation, but this is often not how practicing teachers work, which exerts its own pull on teacher education. The specific implementations of programs take different forms at the universities that offer mathematics teacher education. In order to provide a comprehensive yet meaningful introduction to both the current system and current practices, we

[a] Stockholm University; [b] Linnaeus University; [c] KTH, Royal Institute of Technology; [d] University of Gothenburg; [e] Karlstad University; [f] Umeå University

International Perspectives on Mathematics Teacher Education, pages 9–48.

describe the overall organization of Swedish mathematics teacher education, and then offer short cases of implemented programs. To ensure inclusivity, the various parts are written by mathematics educators from the respective institutions. In this way, both variation across mathematics teacher education for different grade levels and variation across different institutions working within the same national directives can be distinguished. Issues such as the academization of teacher education are problematized, as are other forces that constitute the crosscurrents in Swedish mathematics teacher education.

INTRODUCTION

At the time of writing this chapter in 2019, Sweden's population is around 10 million. Therefore, it is reasonable to ask what the Swedish "case" may have to offer an international audience. In some ways, the Swedish situation is unusual, in other ways not. Though Swedish learners[1] perform among the top countries on the International Civic and Citizenship Education Study, have good reading skills, and generally like school (Sumpter, 2018), the results for natural science and mathematics were on a downwards trend, which only turned with the 2012 PISA results to end up around the Organization for Economic Cooperation and Development (OECD) average (Skolverket, 2019). Mathematics teaching is dominated by teacher-led introductions followed by learners' seated task work, and many teachers adhere to the textbook (Boesen et al., 2014). Yet there is a national requirement to base both content and teaching on "scientific grounds" and "tested experience," which goes hand in hand with a broad research base in what here is called "didaktik" (Rønning, 2019). Teachers are also supposed to work collaboratively—but with little time to do so as the space within which teachers operate is restricted (Sumpter, 2018).

The national requirement informs teacher education in several ways—one of which is the stipulation that student teachers complete one or two independent research projects. The government finances school-university collaborations in order to develop teaching, though these projects may be difficult to implement in practice; for instance, studies indicate that tasks other than teaching occupy a significantly larger part of teachers' time than previously (Ellegård & Vrotsou, 2013; Forssell & Ivarsson Westerberg, 2014). There are national guidelines for teacher education programs in terms of composition, duration, outcomes, and entry requirements, yet there is substantial academic freedom in the design of the local configuration of programs.

Below, we provide a short overview of the school system, followed by a description of teacher education as it is now, and a detour into some history. We go into more depth around the current national curriculum and specified outcomes and address the independent research and the practicum[2] components, which all

---

[1] Throughout this chapter, the word "learners" refers to children in preschool through lower secondary school, and the word "student" refers to prospective teachers.
[2] In the remainder of the text, we refer to any period that student teachers spend in schools as "practicum." The school-based mentoring teachers will be referred to simply as "mentors," and university-based mentors will be referred to as "lecturers."

programs have in common. In the second part of the chapter, we provide examples of specific implementations of programs for preschool, lower primary, upper primary, lower secondary, and upper secondary mathematics teacher education.

THE SWEDISH SCHOOL SYSTEM

When schooling for all became law in 1842, schools for children of the less affluent had already existed in places, but the new law gave parishes[3] five years to organize venues and hire teachers. Hence, the responsibility for teaching the broader population to read changed from church to school. However, it was a parallel system, where more wealthy families sent their children to grammar schools with a more "academic" curriculum. Gradually, this changed. Building a public-school system equal for all was initiated in the 1950s, and finally written into law in 1972, under a social-democratic government. During the mid-1990s, the bipartisan political collaboration around schooling came to an end, and several reforms were made over a short period of time (Imsen, et al., 2017). This continued during the beginning of the 21st century.

The grading system also changed over time. Between 1897 and 1962, learners were assessed by how well they mastered the curriculum, but there were few guidelines for teachers. Between 1962 and 1994, grades were determined relative to a normal distribution of performance. In 1994, grades again became tied to the outcomes of the curriculum, but with specific criteria for each grade (Melén & Wedman, n.d.).

The introduction in 1992 of the freedom to choose a school instead of being bound to attend the neighborhood school is an example of a political decision with significant impact, as research strongly suggests that it led to reduced sociodemographic diversity among schools (Åstrand, 2016; Dahlstedt et al., 2019; Spaiser et al., 2018). It even became possible for organizations and businesses to run schools (Dahlstedt et al., 2019). Moving authority for schools from the national to the municipal level changed power relations throughout the school system. Although there is still a national curriculum and national laws regulating the school system, the economic responsibility lies at the municipal level. Teachers, who were previously nation-state employees, became municipal employees, paving the way for heterogeneity in employment conditions and salaries. Now, only universities are nationally run.

Combined with the changing demographics of Sweden due to immigration (Sweden is the EU country that has received the highest number of refugees per capita in recent years), the free school choice and changes to authority structures have challenged the ideas underlying a unified school for all. Naturally, these structural and ideological changes to schooling affect the education that prepares teachers for the schools of today and the future.

---

[3] Historically, Sweden was organized into parishes, which were subdivisions of the Church of Sweden.

The most recent curricular change was the addition of computer (and related) programming to the mathematics curriculum for all grades.[4] After an open letter in a business paper, and an acknowledgment from the minister of education that he had been in conversations with Information and Communication Technology (ICT) and tele-communication businesses,[5] some suggest that this change came about due to industry pressure. In the open letter, the founders of a young company wrote:

> The demand for programmers and developers is enormous and therefore we must plainly educate more. ... we need to get programming in early in primary school, so that we nurture existing talent as well as prevent the loss of female programmers. Woodwork is still mandatory in Swedish schools, but programming is not.[6]

To a fair extent, programming is required to be integrated into mathematics teaching, to the lament of some and the delight of others. On the one hand, programming is linked to mathematics through abstractions, algorithms, logic, and modeling (Niemelä & Helevirta, 2017; Wing, 2006). It can therefore be argued that it has obvious links to the content. On the other hand, skillful teachers are needed for (mathematics) teaching to take place through inquiry-based methods (see Kirschner et al., 2006), and adding programming to the mix does not make this less complicated. Therefore, some feel that this curricular change not only adds to an already crowded curriculum, but also complicates the pedagogy.

Today, the school system is divided into preschool, preschool/reception class, compulsory school (i.e., primary school and lower secondary school), upper secondary school (called gymnasium), and further education (i.e., university, university colleges, and higher vocational education)—see Table 2.1.[7] Grades kindergarten–9 are compulsory, and roughly 91% of 24–35 year olds have attained at least an upper secondary education qualification (OECD, 2012). Mathematics teachers for grades 7–12 are specialists, and teachers for the younger grades often teach across most subjects of the curriculum. Teaching generally is conducted in Swedish. Other types of educational activities are also offered, including a track for children and adults with "learning disabilities," an extensive range of adult education including the so-called "folk high schools,"[8] and widespread leisure-time centers for after-school activities.[9]

---

[4] In the early grades, these activities include "programming" fellow learners or toys; i.e., children learn to plan sequences of actions and loops.

[5] https://digital.di.se/artikel/spotify-hyllar-satsningen-pa-programmering-fran-forsta-klass

[6] https://www.va.se/nyheter/2016/04/12/oppet-brev-fran-spotify/

[7] A visual overview of the Swedish school system can be found at https://www.skolverket.se/for-dig-som-ar.../elev-eller-foralder/skolans-organisation/fran-forskola-till-universitet---sa-har-hanger-det-ihop.

[8] Folk High Schools are a Nordic model for adult education. Often, the participants reside at the school for a period. Generally, the education is not part of the formal education system. However, in modern times, some folk high school programs can function as an alternative to upper secondary school.

[9] Leisure-time centers provide a space for younger learners (typically up to the age of 12 or the end of grade 6) to spend time after school while parents are still at work. Various leisure activities, sports,

TABLE 2.1.   The Swedish School System

Swedish School System	Grades	Ages
Preschool		1–5
Preschool Class/Kindergarten	0	6
Lower Primary	1–3	7–9
Upper Primary	4–6	10–12
Lower Secondary	7–9	13–15
Upper Secondary	10–12	16–18

Note: We use "Kindergarten" (abbreviated K) to refer to the Swedish preschool class, the last year before formal schooling. This is equivalent to the reception grade (grade R) in other contexts.

Preschool became part of the formal school system in 1998. Since then, municipalities have been responsible for ensuring that every child between one and six years of age is offered early childhood education. Attendance is, however, voluntary. There are both public and independent preschools, where the latter can be run as a parental cooperative, by a foundation, or by a limited-liability company. Municipalities are responsible for ensuring quality and safety in both public and independent preschools (National Agency for Education, 2019).

Similar to international trends, there is a growing consensus in Swedish policy and research that early education is important for children's short- and long-term development. Therefore, preschool fees are low and related to parents' income, and from the year children turn three they are entitled to 525 free hours in preschool per year (three hours a day). Almost all Swedish children attend preschool: in 2017, 84% of one-to-five-year-olds attended preschool; among four-to-five-year-olds, attendance was close to 95% (SKL, 2018a). This is high by international standards but similar to other Nordic countries (Reikerås et al., 2012). Since the autumn of 2018, it has been obligatory for children to attend kindergarten. It is a maximum of 6 hours per day; the activities are dominated by play, but with transitioning into schooling.

From kindergarten to university, education is free. Learners in upper secondary education (grades 10–12) and university students get national study support. Although the amount is not enough to live on, it reduces the need for loans or extensive work responsibilities during study years and grants young people more independence. Free school lunch is provided for all learners up to grade 12.

There are 18 common programs or tracks at the upper secondary level, of which 12 are directed towards vocations (for instance, the care and healthcare program), and six explicitly prepare learners for higher education (for instance, the economics program). The upper secondary schools can apply to offer special-

and music are often offered, and learners are provided with an afternoon snack or even a meal.

ized tracks, such as combined arts and science, the international baccalaureate, or a mathematics-focused track.

TEACHER EDUCATION IN SWEDEN

Paralleling the school system, teacher education is divided among the following programs: (1) preschool teacher; (2) primary teacher for grades K–3; (3) upper primary teacher for grades 4–6; (4) lower secondary teacher for grades 7–9; (5) upper secondary teacher for grades 10–12; (6) folk high school teacher; (7) leisure-time center teacher; (8) vocational teacher; and (9) special needs teacher. In this chapter, we only look at (1)–(5). Only one university has a specific program for folk high school teachers—a one-year program. The program for leisure-time center teachers is three years, and the graduates are qualified to teach a practical or an esthetic subject (e.g., physical education, art, or home economics). Vocational teachers need experience in the vocation they want to teach, and supplement this with an additional 18 months of teacher education. They will teach in the vocational streams of upper secondary school. Special needs teacher education is an Honor's/Master's program and a completed teacher degree is an entry requirement.

The present version of Swedish teacher education was introduced in 2011. This reform entailed more detailed regulations at a national level and focused on more specialization of teachers for different age groups. All programs are overseen by national authorities, which ensures that entry into and the overall structure of the programs are the same across universities. This means the same weight of content and pedagogical content courses, school practicum, and independent research projects across all institutions (see more below). It also means the same outcomes on the program level across all institutions. This naturally implies some degree of national coordination and political/administrative oversight—including that student influence is enabled across the system.[10] Content and delivery of courses vary across institutions, which are also free to formulate specific course outcomes and decide on assessments. For access into a mathematics teacher education program, an applicant must have completed upper secondary school with certain mathematics courses, depending on the grade level at which the program is directed.

Sweden has double the number of universities per million inhabitants as the USA. Thirty-three higher education institutions offer teacher education, although a few of these are specialized toward, for example, music, art, or physical education. Twenty-four of these have recognized mathematics teacher education programs, and most institutions offer these programs across all levels. Previously, teacher education took place in separate colleges, but was assimilated into universities around fifty years ago. The independent research component was included in 1993, in a revision of the programs aiming to give teacher education a scientific

---

[10] By law, students have the right to be represented during decision making and preparation for decisions at all levels.

base (Råde, 2014), but the academization process started in the 1970s (Karlsudd, 2018).

Mathematics teacher educators are mostly former mathematics schoolteachers. The extent to which they have acquired additional research degrees varies among institutions. Posts at associate professor level and above require a Ph.D. and are often advertised both in Swedish and English, attracting some international applicants. Long term, the national aim is that all teacher educators should have a research degree, reflected in the relative frequency of Ph.D. graduates becoming an indicator in overviews of the program implementations (Bejerot et al., 2018). This is contentious, as it is at times seen to be downplaying the experiences from teaching in schools. A recent survey of 1,554 upper secondary teachers' feelings of preparedness after their education plays into the debate, as the study found that the aspect of practical professional knowledge had deteriorated over time, yet other areas had improved (Bejerot et al., 2018).

**Vignette 2.1. An Example of a Collaboration
Between a University and a Secondary School**

Student teachers Agnes and Oscar had been working on two mathematics tasks together with three 17-year-old learners, and conducted a follow-up interview with the learners. Agnes and Oscar summed up the interview: "Motivation was not strong and mathematics was perceived as difficult. The most engaging thing in mathematics is when you understand it, when you get the 'aha-experience.' It is worth noticing that the learners felt more engaged with the first task where they could recognize a pattern. They also expressed how task 1 was more rewarding. One of them said 'It seemed difficult at the start but got easier when we were asked to write down our ideas.' Task 2, which was based on remembering definitions and concepts, made them uneasy and they felt they could not demonstrate their potential."

This experience was part of a collaboration between a university and an upper secondary school. The student teachers were invited to work with a mathematics teacher, Richard, who had an ongoing experiment using oral rather than written mathematics assessment tests. The learners in Richard's classroom were in a social studies program, where several of them did not value mathematics, at times demonstrating resistance towards and even fear of the subject. For the participating student teachers, working with Richard was part of a mathematics education course where they read about norms and affect in relation to mathematics, and it made them reflect, among other things, on the consequences of choosing different tasks.

The other participants learned too. The collaborating school had extensive practicum mentoring experience, yet felt that designing teaching with several adults in the classroom generated new opportunities. The lecturers on the campus course saw an opportunity for engaging more closely with mentor teachers around the coursework of students.

Teacher education accounts for a quarter of the intake to all professional university programs; despite this, there is a serious shortage of teachers in Sweden (SKL, 2018b). A recent prognosis anticipates a 15% increase in learners between 2017 and 2027, resulting in a need for 600 new preschools and 300 new schools (SKL, 2018b). This has also led to increases in teacher salaries, making it a challenge to attract new teacher educators to the universities (SKL, 2018b).

Graduating from a teacher education program is sufficient to apply for the official certification that allows the graduate to teach. Certification is only denied if there are compromising circumstances, such as mental health issues or a serious criminal record. Induction is up to schools. A qualification from any institution is valid all over Sweden (and the European Union, although some additional individual country requirements may exist, such as language competencies). When other professional education programs in Europe were aligned in 2007, the alignment did not apply to teacher education, but later even these programs were adjusted (see Swedish Council for Higher Education, 2019).

The Swedish government provides reasonable funding opportunities for collaboration between schools and universities, serving the dual purpose of supporting the research foundation of practices in schools, and facilitating better teacher education. This can take many forms. The narrative in Vignette 2.1 is one case of such a collaboration.

VISIONS FOR MATHEMATICS
TEACHERS THROUGH HISTORY

This section explores the vision for mathematics teacher education conveyed by the Swedish government. Throughout the last century, the government solicited several national inquiries into teacher education, and this section discusses the reports from 1965 (*Statens offentliga utredningar* (SOU) [Swedish Government Official Reports], 1965:29), 1978 (SOU, 1978:86), 1999 (SOU, 1999:63), and 2008 (SOU, 2008:109). These reports are all concerned with two problems: the lack of mathematics teachers, and the lack of quality in mathematics teacher education. The 1965 report made clear that these are not independent of each other, but rather two aspects of the same problem (SOU, 1965:29). This constitutes a dilemma: increasing the demands of the program may improve quality but will likely lead to fewer applications, and vice versa.

Sometimes the reports comment on particular challenges with mathematics teaching. One particularity is the changing importance assigned to mathematics in the reports — as an important skill but also for developing a taste for beauty (SOU, 1965:29), as necessary for personal development (SOU 1978:86), or as part of the basic knowledge for school and life (SOU, 2008:109). Another particularity is the changing attention given to mathematics learners' difficulties. Knowledge about difficulties in mathematics was proposed to be included in teacher education in 1978, whereas the 1999 report assigned it to special needs education. The 2008 report focuses on the psychological features of learners, and Sjöberg

(2010) describes an emerging dichotomization between the normal and the deficient learner. At the same time, teachers are made accountable for the success of all learners in line with a language of economy, focusing on effectiveness, goal-achievement, and results (Mickwitz, 2015; Sjöberg, 2010).

This, of course, is reflected in the visions for mathematics teachers in the reports. One vision that recurs is about the knowledgeable teacher — knowledgeable about the content to be taught. Another is the vision about good mathematics teaching, albeit viewed rather differently over time. As early as 1893, a government report described teaching as the "ability to easily and clearly communicate content" (Utsedde Kommitterade, 1893, p. 39). When the school system expanded during the 1950s, teaching needed to reach out to new groups of learners, and the suggestion in the 1965 report was for individualist teaching of mathematics, with workbooks learners could follow at their own pace. In the 1999 report, the focus shifted to the learners' interest in mathematics, whereas the focus on grading and assessment in the 2008 report (Mickwitz, 2015) reinforced a vision of individual achievement in mathematics.

The Scandinavian school systems share values of democracy and equity (Rønning, 2019), and those values are visible in expectations of teacher education. Both national and global equity and solidarity are foregrounded in curricular and guidance documents. The values are reinforced in demands for a socially broadened recruitment of teachers, as well as in demands for the teachers to foster such values in learners (Sjöberg, 2010). However, a shift takes place in the 2008 report. Although schools are still expected to practice equity in relation to mathematics learners, the increasing diversity among student teachers is now seen as problematic, as yet another sign of poor quality. In this report, implicitly, the ideal student teacher seems to be a high-achieving male from a favorable background.

Studies of these reports all conclude that the 2008 report was different (Mickwitz, 2015; Österling, in preparation; Sjöberg, 2010). Whereas the previous reports acknowledged the challenges of providing schools with teachers in large numbers while keeping up quality in teacher education, the 2008 report takes its starting point from the shock that followed the Swedish PISA results in 2006 (see Sollerman, 2019). Within a discourse of effectiveness and goal-orientation, this so-called failure of learners is attributed to the failure of mathematics teachers and thereby the failure of mathematics teacher education. Following this logic, teacher education must improve achievement through assessments of teachers and learners. This perspective is not always compatible with the historically strong vision of a teacher education aiming for equity, democracy, and interest in knowledge.

In summary, numerous changes over the past few decades have been proposed by the Swedish government, leading to shifted visions for mathematics teacher education. In Vignette 2.2, an experienced mathematics teacher educator reflects on these changes over time.

**Vignette 2.2. An Experienced University Professor in
Mathematics Education Reflects on the Changes Over Time**

When I became a teacher in the spring of 1971, I had spent three years at the university and had studied at least one year each of mathematics and chemistry—without any connection to teaching. I then spent one year at the College of Education in Stockholm studying teaching methods and pedagogy, with 9 weeks of practicum in different schools. The second semester included a (salaried!) half-time practice period at an upper secondary school. It was important that subject theory, teaching methods, and pedagogy would be translated into school teaching. The program permitted teaching at both the secondary and upper secondary school. Those who were to become primary school teachers only attended the College of Education. It was difficult to get into teacher education because competition was so stiff.

What is the situation now? It is easy to get into teacher education (but this does not mean that the applicants are weak). The program looks very different and is longer but with less teaching time. (When the Stockholm College of Education was founded in 1956, the students complained about the extensive practicum component in their education.) During the entire education of primary school teachers, mathematics is taught integrally with math education. In the 1970s, we received teacher-skill grades, with three levels of passing. Criteria for different grade levels were not explicit; it was the lecturers' perceptions of the desired teacher which determined the grade. A lecturer in the methods course put it this way: "I look at the extent to which you firstly have a happy and positive climate in the class, secondly that you explain so that the learners understand, and thirdly that you almost cover the content." Now the entire program is divided into separate courses and there are pre-determined grading criteria for each course to which the students must have access.

Who used to teach teacher education? They were hand-picked, skilled teachers from schools who were also considered to be skilled student teacher supervisors. Thus, one may say that it was a form of nepotism. Nowadays, employment is advertised and application is competitive.

In the past the most central thing was that the program produced teachers who could translate relevant knowledge in such a way that all learners learned and developed. The verified experience was strongly emphasized, while the scientific basis was completely left in the shade. Now, every student teacher completes at least one research study that demonstrates competence in engaging scientific theories, existing research, and application of research methods. According to evaluations of teacher education, this requirement has proved difficult for the higher education institutions to fulfill, even though teacher education departments now have more scientifically educated teaching staff.

How should we find a balance between student teachers developing into proficient teachers in the classroom, where they can provide their learners with a stimulating environment for learning and personal development, while ensuring that what they do rests on scientific foundation and tested experience?

THE CURRENT NATIONAL GUIDELINES AND
GOALS FOR TEACHER EDUCATION

The composition of programs is determined by the Qualifications Ordinance (Swedish Council for Higher Education, 1993/2018), and varies depending on grade level, but all contain core education subjects, practicum periods, and specializations, which include one or two independent research theses (see Table 2.2).

Next, we address the core education component, the practicum component, and the independent research component, as these are common to all programs. Later in the chapter we go into details with the mathematics and, in particular, the mathematics education components in the discussions of the separate programs. This includes the "specialist education" component of the preschool teacher education program.

Core Education

The core education courses cover the same aspects across programs, namely history of the school system and its core values, curriculum theory, theory of

TABLE 2.2. ECTS Points[a] of Various Components in Swedish Teacher Education Post-Secondary Programs

	B.A. in Preschool Education	Professional Master's in Primary Ed.[b] (for grades K–3 or 4–6)	Professional Master's (M.A./M.Sc.) in Sec./Upper Sec. Ed. (grades 7–9 or 10–12)
Core education	60	60	60
Specialized education	105 (early years)	—	—
Subject knowledge (mathematics, physics, etc.) Subject education (mathematics education, science education, etc.)	Requirements given at the institutional level, but included in "specialised education"	120	120–150/180–210 (depending on subject combination)
Placement / subject specific + practicum	30	30/15+15	15+15
Independent research	15	30	30
Total (length of time beyond upper secondary education)	210 (3.5 years)	240 (4 years)	240–330 (4–5.5 years)

[a] ECTS stands for the European Credit Transfer and Accumulation System. 30 ECTS points correspond to one semester of full-time studies. It is customary to omit "points."
[b] The Professional Master's is a one-year degree after the bachelor's degree in the European system, sometimes called the "advanced degree." It includes a half semester research component.

knowledge, research methodology, learning theory, assessment and marking, social relations, leadership, and evaluation and development processes. The outcomes for all programs are divided into three parts focusing on knowledge, skills/application of knowledge, and values. We summarize these in the sections below. In later sections, we exemplify the selection of content of the mathematics and mathematics education courses in implementations of the various programs at different institutions.

Knowledge Components

A range of knowledge components are listed. The content knowledge requirement increases with level of schooling for which the student teacher prepares, up to expecting secondary teachers to have both an overview of the discipline and specialized knowledge in a part of the discipline (e.g., linear algebra, statistical inference, vector geometry—typically the topic the student chooses for his or her Bachelor-level thesis in the subject). Students must know subject-related education content (i.e., mathematics education), and have knowledge of learners' development, learning, and needs. As Swedish education is expected to be based on research, all students are required to display understanding of the theory of knowledge, research methods, and the relationship between the discipline, research, and professional practice, and to have an awareness of or insight into current research and development work. They must also understand social relationships, know conflict management, and demonstrate leadership. They must have knowledge of the school system and its history, and of relevant policies and curricula. All but the preschool student teachers must know about assessment, and for grades 4–12 also about marking.[11] Future teachers for preschool through grade 6 must "demonstrate knowledge of practical and aesthetic learning processes" (Swedish Council for Higher Education, 1993/2018) and know about the acquisition and development of reading, writing, and mathematical skills. Finally, preschool through grade 3 teachers are required to know about children's communicative and linguistic development.

Required Skills and Applications of Knowledge

The required skills and applications of knowledge are almost the same across all programs. Graduating students must be able to apply their general and subject-related educational knowledge and methodology; build on learners' knowledge and experiences to facilitate learning for all; be able to plan, implement, evaluate, and develop educational processes; be able to identify and together with others deal with special needs; be able to observe, document, analyze, and communicate about the development and learning of the learners; be able to convey human rights, equal rights, democratic values, and prevent discrimination and harass-

---

[11] No grades are given until grade 6, which explains why it is not necessary for student teachers for grades K–3 to engage in this component as part of their education.

ment; have good communication skills with listening, speaking, and writing; and be able to use digital aids critically. The future preschool through grade 3 teachers must be able to inspire learning and meaningful leisure (play), and future teachers for the higher grades must have specialized capacity to create conducive learning environments. Finally, all future teachers must be able to learn from systematic and critical reflections on experiences and research, acquire skills from practice, and contribute to their own professional development as well as the formation of knowledge in the field of their professional practice (and for student teachers of secondary school also to mathematics or mathematics education—a tall order).

Value and Aptitude Outcomes

The value and aptitude outcomes concern self-awareness, empathy, professionalism, identification of one's own need for learning and development, and "the capacity to construct assessments in educational processes on the basis of relevant scientific, social and ethical aspects with particular respect for human rights, especially children's rights according to the Convention on the Rights of the Child, and sustainable development" (Swedish Council for Higher Education, 1993/2018). Together, this set of outcomes reflects an image of the desired teacher as one who can apply solid educational and content knowledge in making informed judgments in practice, is imbued with egalitarian and democratic values, and is capable of not only facilitating the teacher's own development, but also contributing to the overall profession or even the research field. This has been characterized as the ideal being *the constantly improving teacher*, according to Christiansen et al. (2019).

It is clear from these outcomes that the education of teachers must include theory and practice and relate these to each other. Exactly how this is implemented is up to each university, making it challenging for student teachers to transfer to another university. In Vignette 2.3, some comments from student teachers are presented regarding theory and research.

As mentioned, teaching in Sweden is meant to be based on "scientific ground," and this shows up in the outcomes and structure of the teacher education programs. Not only is a comprehensive knowledge base required, but the prospective teachers are expected to contribute to knowledge generation. This becomes very real for the student teachers in the final research project that caps their education. This is addressed in the next section.

The Independent Research Component

An independent research component is included in teacher education at all levels (Swedish Council for Higher Education, 1993/2018). Each teacher education student is required to complete one or two research projects for a total of 30 ECTS (European Credit Transfer and Accumulation System points). In most universities this is done as two projects, one after about three years of the program, and one at the end. The projects are often small empirical studies using classroom

**Vignette 2.3.   Some Student Teachers Comment on
Theory and Research in Their Education
(obtained from interviews conducted in the TRACE project)***

"They say that schooling should have a scientific base, and to do that, we must at least be able to understand research. If you take a random article in mathematics education but you have never heard about any theories of learning, you won't understand anything."

"I have strong faith in a classroom focus on problem solving, and then without searching for it, you get to Vygotsky. I want a living classroom climate, I don't want silence, I don't like it. Sometimes it may be so, again, vary the teaching. If you make people talk then you will have learners who show their competencies in other ways as well."

"... today, [someone] asked me, how can we use Brousseau's idea about delegating the task to learners until they have solved it. Well, I did that during my first practicum—here, now you get this sort of task—but I did not understand that that is what I did. [...] What I think is that I get much clearer ways of varying my pedagogical approach, in a way I have words for. Yes but I can do this. Perhaps I should vary that."

*https://www.mnd.su.se/english/research/mathematics-education/research-projects/trace

---

**Vignette 2.4.   Two Examples of Independent
Research Projects by Future Teachers**

Elin Gawell: *"But that's not really how the Swedes do math": Immigrant high school learners' conceptions of school mathematics and math education.* Stockholm University (Gawell, 2017). In the project, six upper secondary students who had attended primary school in another country were interviewed, and the interviews analyzed using, amongst others, a framework of socio-mathematical norms. From the abstract: Results show that mathematical methods as well as the type of mathematics that is emphasized differ between countries. Especially, there were substantial differences regarding emphasis on problem solving and what is regarded as a good mathematical solution.

Daniel Vulic: *An analysis of the introduction of the equal sign in Swedish textbooks.* Stockholm University (Vulic, 2017). Grade 1 mathematics textbooks were analyzed with respect to the types of equality expressions used. In addition, teacher guides were analyzed for suggested strategies for introducing the equal sign. From the abstract: The result shows great differences between the analyzed books. There were few non-standard equations in the learner workbooks following the introduction section of the equal sign and the support between different teacher guides varied considerably.

observations, surveys, or interviews with learners or teachers. Two examples of independent research projects by future teachers are shown in Vignette 2.4.

Research questions could address teachers' ways of talking about assessment, their teaching of specific content, or how they work with multilingual learners. Focusing on learners, questions could concern their understanding of specific concepts, how they think about the subject, how they are assessed, or how they engage with homework. Fairness issues are often raised in the research questions. The formulation of the research questions is of great importance to the outcome. Karlsudd (2018) studied 58 projects where most of the research questions were descriptive, and his main conclusion was that a well formulated research question is most often connected to a higher grade.

Learning goals for the independent research component cover the planning, implementation, and reporting on a study that addresses the research question as well as a discussion of the results in relation to previous research and the teaching profession. Hence, teacher education in Sweden today has the goal of preparing teachers for research, wherefore the goal of the independent research component is to provide knowledge about research methods, foster a critical and scientific approach, and add knowledge to professional practice.

As mentioned previously, the independent research component has been a requirement in Swedish teacher education since 1993 (Råde, 2014), as a culmination of a movement towards a more scientifically based education (Karlsudd, 2018). Previous reports have noted that the individual research component focuses more on the scientific dimensions than on problems relevant for the profession (Gabrielsson, 2005), creating a tension between the professional and the scientific part of teacher education (Fransson, 2009; Karlsudd, 2018). This tension, between professional-practical knowledge and the demands of a scientific report, has been identified in other countries as well (e.g., Meeus et al., 2004; Reis-Jorge, 2007), though Finland appears to be an exception.[12]

The results of the rapid "academization" of teacher education has not only meant that many prospective and practicing teachers feel that they have lost their profession (Calander, 2008), but it also appears to have created a Catch-22 moment: the identified lack of scientific knowledge in teacher education leads to the foregrounding of scientific learning goals, but the students struggle to fulfill the goals. As a result, there is little or no contribution to the professional practice as originally intended by the Ministry of Education and Research. This is a challenge for Swedish teacher education.

Another setting for potential tensions between professional-practical knowledge and the scientific basis for teaching occurs in the practicum component. We now turn to a discussion of the practicum component of teacher education programs in Sweden.

---

[12] Finnish teacher education has a long tradition of higher academic education, including writing a master's thesis (Maaranen, 2010). In a comparison with teachers from Norway, Finnish teachers exhibited a more analytical and specific professional language when discussing educational matters (Wågsås Afdal & Nerland, 2014). Here, it appears that the standards of science and profession meet.

The Practicum Component

Since 1999, practicum has been referred to as a school-based *education* (verksamhetsförlagd utbildning, VFU), where *education* signals the aims of not only practicing teaching skills, but also giving future teachers opportunities to integrate the school-based component into their entire teacher education (SOU, 1999:63). Each university was entrusted with planning for practicum, but the time and quality of practicum placements were found to differ. Thus, the next report suggested specifying credits for practicum courses, and ensuring quality by regulating learning outcomes (SOU, 2008:109). This report informed the 2011 reform of teacher education. As a result, a tension arose between practicum as an integral component of programs, and separating goals and credits for practicum into a separate course. This section discusses how such tensions in the political directing of practicum are handled by the different institutions.

All Swedish teacher education programs contain at least 20 weeks of practicum (30 ECTS). Still, practicum is organized differently at various universities. We conducted a brief survey, which helped us describe some variations.[13] Most commonly, the 20 weeks are distributed over three or four separate courses as blocks for the duration of the program. Some universities argue for the importance of future teachers continuously integrating coursework and placements, but only two of the six universities that replied have students spend time at schools more frequently than four times over their education.

Although universities make such independent decisions, practicum is regulated nationally on the level of assessment. Since 2014, practicum courses were the only university courses mandated at the national level to be graded with a multiple-level scale; that is, pass or fail was not sufficient (see Swedish Council for Higher Education, 1993/2018).[14] Student unions protested in vain, and this issue is a significant indication of how teacher education has become a battering ram in the political arena.

Apart from grading scales, decisions about criteria and the means of assessing practicum were left to the different universities. In the case of Stockholm University, assessment criteria were developed from the national outcomes. This entails a strong alignment with policy, and at the same time a potential detachment from previous teacher education experiences and research on practicum. Criteria for practicum are generally aimed at reflecting a progression in competencies (Christiansen et al., 2019). For instance, the practicum courses at Stockholm University are built on the idea of future teachers learning to become progressively more independent. In the first practicum, future teachers are expected to teach a few lessons prepared with their mentor. In the final practicum, the demands of planning and teaching independently are higher.

---

[13] A special thanks to Gothenburg, Halmstad, Jönköping, Linnaeus, Mid-Sweden, and Uppsala Universities for their contributions to this section.

[14] Other courses may use multiple-level scales or pass/fail, depending on university requirements.

Assessing teaching is a complex task, and relative to the prevailing vision of good teaching (Christiansen et al., 2019). Assessing practicum is consequently equally complex and is related to the visions of good teaching held by both the mentor and the student as well as the visions conveyed by lecturers of university courses.

Student reflections are used as additional grounds for assessment, as students are expected to reflect on their practicum in relation to their campus-based studies. Studies of such reflections suggest frequent gaps between theoretical and practical learning (Emsheimer & Silva, 2011). One example of this is given in Vignette 2.5, which is the narrative of a student teacher's commentary on the mentor-mentee relationship during her practicum.

Research reflects an on-going search for valid instruments and criteria for assessment of practicum (e.g., Rusznyak, 2012; Wyatt-Smith & Klenowski, 2013). In Sweden, the responsibility for examining lies with the university, often in consultation with the mentors. The instruments supporting learning and assessment of practice are based on previous experiences and research and are constantly submitted to both development and research. Examples of such instruments are teaching portfolios (Dysthe & Engelsen, 2004), application tasks (Christiansen et al., 2018), "triadic conferences" between a student, mentor, and lecturer (Jons, 2019), or checklists and commentaries for lesson observations (Christiansen et al., 2019). Despite this continuity, the changes in government policy also create ruptures in the development of instruments.

**Vignette 2.5.   A Student Teacher's Commentary
on the Mentor-Mentee Relationship in Her Practicum
(based on an interview with a mentee, conducted in the TRACE project)**

Greta, a mathematician who broke off her doctoral studies and is now studying to be a secondary school mathematics teacher, makes plans for the coming four weeks with her mentor, Sven. Greta is somewhat provoked since Sven plans to spend all this time on units. She wonders when they can engage in mathematics rather than such simple stuff. However, Sven is the mentor and they begin to work with units and how to convert between, for example, millimeters and decimeters. Slowly, it becomes obvious to Greta how converting units really can be a struggle for learners. She realizes that, as a teacher, she has to teach simple content so the learners can use it in more advanced problems later. However, she feels that it should be possible to work more with the relationships between the units for different quantities, as well as the positional system of numbers. She feels that there is a difference between how mathematics is described in her teacher education courses and how it is practiced in her practicum class. Mathematics in teacher education focuses on relations, while during practicum it is more about rules. In Greta's opinion, mathematics is more about relationships and that is something she would like to change in her future mathematics teaching.

One such rupture was found when comparing the application tasks from secondary mathematics teachers' practicum before the 2011 reform and after (Österling, in preparation). Before 2011, when practicum was integrated into courses of mathematics education, students were asked to do application tasks related to the course content, for example, to select and teach around a mathematical problem from the course book on problem solving and evaluate how learners solved the problem. After the 2011 reform, practicum was given as a separate course, and in the case of Stockholm University, the tasks for the new online teaching portfolio were developed without specific references to courses. Instead, student teachers were asked to formulate their own goals and reflect on their learning. The final task asked students to identify strengths and shortcomings and reflect on how to develop further professionally—which is similar to the national goal that teachers must "demonstrate the capacity to identify the need for further knowledge and to develop his or her own skills in pedagogical practice" (Swedish Council for Higher Education, 1993/2018). One main difference is the lack of connections to mathematics or mathematics education in the recent tasks. Furthermore, the responsibility for identifying the characteristics of good teaching was delegated to the student teachers. Hence, the separate practicum led to reduced or unclear expectations on students to integrate knowledge from courses with the practicum experience.

On one hand, such responsibility assumes the necessity for teachers to reflect on their actions (see Korthagen & Vasalos, 2005; Schön, 1987). This, together with specialized knowledge, is the basis for decisions and professional judgments in teaching (Rusznyak & Bertram, 2015). On the other hand, reflection as self-inquiry is a common practice in mathematics teacher education (see e.g., Adler & Davis, 2006). Compared to reflection on actions, there are different logics underlying such self-reflection. Korthagen and Vasalos (2005) have described it as "core reflection," (p. 53) where teachers need to reflect on their identity and mission in relation to the classroom. A different logic related to self-inquiry is echoed in the regimes of accountability (Biesta, 2013), where a top-down goal-orientation requires teachers to evaluate themselves in relation to external goals. The national goal described above, where teachers are required to "identify their need for further knowledge ... ," is in line with this top-down logic, where a self-reflection becomes part of an accountability regime. It is akin to a form of confession of sins committed against the vision of a good teacher (Österling, 2020).

When practicum is organized as independent courses without explicit links to mathematics education courses, there is a risk that reflection as self-inquiry dominates, at the cost of reflections on actions that inform professional judgment (Österling, 2020). What appears to be an organisational move turns out to potentially undermine the visions of integrated teacher education, which it is then up to institutions to keep alive.

PRESCHOOL AND PRESCHOOL TEACHER EDUCATION

Sweden has a long tradition of early childhood education and care. The first form of an early childhood curriculum was written in 1837 (Eidevald & Engdahl, 2018). The Swedish preschool has a tradition of social pedagogy (Bennett & Tayler, 2006) where care, socialization, and learning are to constitute a coherent whole, but also where teaching should take place (SFS, 2010:800). According to the Education Act, preschool should:

> … ensure that children acquire and develop knowledge and values. It should promote all children's development and learning, and a life-long desire to learn. Education should also convey and establish respect for human rights and the fundamental democratic values on which Swedish society is based. (SFS, 2010:800, §2)

Preschool should take its starting point from what children find interesting, and from their knowledge and experiences already acquired. It should be safe, fun, and offer children opportunities to learn through play and explorative activities. Even though the curriculum encompasses several goals to strive for, there are no specific benchmarks or standards for children to attain (National Agency for Education, 2018).

There are two main types of pedagogues working in Swedish preschools: preschool teachers and so-called "child-minders."[15] The preschool teachers are responsible for the educational content and each preschool must have at least one. Preschool teachers' responsibilities include leadership and the organization of activities that are caring and playful, based on the children's experiences and interests. The aim is to provide each child the conditions for learning and developing as far as possible in relation to the target areas, through activities based on the norms and values described in the curriculum (National Agency for Education, 2018).

Even though preschool has had a national curriculum since 1998 and is part of the school system (SFS, 2010:800), the notion of teaching at this level was only inserted in the curriculum with the 2018 revision. On the one hand, the national curriculum clearly states that preschool practice must be designed to ensure that all children develop their competencies—including mathematical competencies—to their full potential; on the other hand, it has no benchmarks, and states that play "is the basis for development, learning and well-being" and "should have a central place in the education" (National Agency for Education, 2018, p. 8). Hence, it is not obvious how teaching in relation to preschool should be understood, and the relation between play and teaching is debated in the preschool community, in preschool teacher education, and in early education research (see for instance Palmér & Björklund, 2016). Is teaching to be integrated with or separated from children's play? Some voices emphasize children's right to play, undisturbed by adults, for the sake of play itself (for instance Sundsdal & Øksnes, 2015), whereas

---

[15] These "barnskötare" have a shorter secondary vocational education.

others emphasize a consolidation of play and teaching (such as Pramling et al., 2019). The former voices often have philosophical underpinnings and are critical of the schoolification of preschool education, and the latter often highlight preschool as being part of the education system and emphasize teaching. Thus, in line with international trends (Palmér & Björklund, 2016; Pramling et al., 2019) there are different opinions on teaching and learning in preschool, reflected in varying views on preschool education and preschool teacher education.

Preschool teacher education in Sweden has its roots in what is known as the seminar tradition, which can be traced back to the mid-1800s. The intention was to educate teachers to teach children from low social-economic backgrounds (Beach et al., 2014); teacher education was considered vocational, where small groups of students learned the craft of teaching. The primary focus of the seminar tradition was on methodological aspects of teaching as well as on questions of morality. Hence, the seminar tradition usually accentuated the practical aspects of teaching and was primarily based on what is known as practice-oriented knowledge. This changed in the 1950s when the seminar tradition was replaced by teacher colleges, which in turn were incorporated into universities during the 1970s.

Since the teacher education reform in 2011, preschool teacher education comprises 210 ECTS, or 3.5 years full time university study with three integrated components (see Table 2.2). The only regulation concerning mathematics is that by graduation the students have knowledge about the learning of basic mathematics and can demonstrate this knowledge in the field of preschool education. Within these requirements, each university may design the content of their preschool teacher education program implementation. Consequently, the amount of mathematics included in preschool teacher education at different universities varies from 7.5 to 15 ECTS. Course delivery and content are designed differently; some universities have a thematic approach, and others have a more subject-specific approach. For example, in the preschool education program implementation at Stockholm University, the mathematics education course (12 ECTS) focuses on pedagogical documentation and how this is extended as a tool to monitor and challenge learning processes, as well as to reflect on the students' own pedagogical actions. The course considers children's mathematical activities and exploration in their daily lives (e.g., through play). At Linnaeus University, the corresponding course (15 ECTS) is divided into three connected parts, one focused on the students' own mathematical knowledge, one practical component focused on the teaching of mathematics, and one part focused on young children's learning of mathematics. The courses at the two universities are different and based on different theoretical foundations, yet the outcome requirements for the students are the same.

A preschool teacher in Sweden can, besides working in preschool, also work in kindergarten. Kindergarten was implemented in 1998 to facilitate a smooth transition between preschool and primary school and to prepare children for education. Kindergarten was optional in Sweden but became obligatory in 2018. To facilitate a smooth transition, kindergartens are expected to design their school-preparation programs as mostly play-based.

MATHEMATICS EDUCATION FOR LOWER PRIMARY PRESERVICE TEACHERS

The primary teacher education program at the University of Gothenburg lasts eight terms, including two mathematics/mathematics education courses, and mathematical content in courses on communication, practicum, and independent research projects. The two mathematics courses are given in terms 2 and 7. In terms 5 and 8, students are encouraged to focus their independent research projects on mathematics education. This is done by providing examples of quality theses, giving inspiration by presenting relevant articles, and introducing questions and methods from the field of mathematics education research, such as communication in the mathematics classroom.

The program has a strong focus on teaching methods in mathematics education, due to a research tradition centered around classroom research and development of pedagogical content knowledge (Bentley, 2003; Kilborn, 1992; Kilhamn, 2011; Löwing, 2004) and a close collaboration with the *National Center for Mathematics Education* (NCM) over the decades. Further, the focus on teaching methods is a result of efforts towards constructive alignment in mathematics courses, in line with national requirements for strengthening the focus on methods and subject education. Research-based approaches addressed are Realistic Mathematics Education, variation theory, the lesson study approach, and the Theory of Didactical Situations.

The Structure and Content of the Courses

Mathematics is integrated with mathematics education content in the courses. The first course focuses on arithmetic and related pedagogical content knowledge—PCK (Hill & Ball, 2009; Shulman, 1987)—and involves contributions from the Department of Mathematics. In the course, the school curriculum content is connected to university-level mathematics, so that students' own mathematical development is furthered. The second mathematics course focuses on pre-algebra and problem solving. The course centers around grade-relevant problems by pursuing and elaborating on the mathematics needed to work with these problems as well as strategies for teaching problem solving. Topic areas are proportionality, statistics, and probability, as well as combinatorics and algebra (mathematical equalities, functions, equations, and patterns).

The two mathematics education components focus on planning and assessment with the aim of teaching for understanding. In both courses, students analyze video-recorded classroom interaction in mathematics lessons from different countries. Analytical tools draw from the repertoire previously listed, making a clear link between theory and practice. This has been positively received by the students, and there are some indications that it has impacted their teaching or the ways they communicate about teaching and learning.

A typical day contains a lecture in the morning and a follow-up seminar in smaller groups in the afternoon. We arrange workshops about, for instance, textbook analysis and assessment of learner solutions. During the first course, students have a three-week practicum, where they are given the opportunity to use ideas from their coursework in teaching. For instance, they work with models such as the ten frame and the number line and engage learners in mathematical discussions giving learners the opportunity to mathematize. Students are examined orally on their planning and lesson performance.

During their second course, there are two field studies of two weeks each, one in kindergarten and one in grades 1–3. Students work with diagnostic materials and are orally examined on how the results contribute to their planning and teaching. One example is a lesson study, with mathematical equalities and the meaning of the equal sign in focus. The other field study focuses on assessment, where a lesson on problem solving is taught.

In the courses, recent research articles and dissertations, both national and international, are presented and discussed. Students are offered support if they connect their independent research projects to the mathematics education course content, such as the importance of communication in the classroom and models as tools for understanding, textbook and content analysis, and other ongoing mathematics education research at the department.

Mathematical Discussion for Inclusion and Other Pedagogical Approaches

One innovative approach of the program is promoting mathematical discussion as a tool for inclusion, which connects to research at the department, especially a project on communication in mathematics classrooms (KOMMA[16]). A framework of communicative actions for planning and orchestrating classroom discussions was iteratively developed by teachers and teacher educators and tried out with several student cohorts (Kilhamn et al., 2019). The framework builds on the idea that every learner should have equal opportunity to participate in a mathematical discussion. Given the right tools, such as norms, aims, and communicative skills, teachers can make mathematical talks inclusive and instrumental in developing mathematical discourse. During practicum visits, lecturers observe how students implement the framework and discuss it. It is noticeable how the use of lesson plans informed by this framework contributes to conceptual clarity and sharpens the focus of lessons.

---

[16] The name of the project, KOMMA, represents both competence and communication (using the Swedish spelling for these words), together with MAthematics.

Students' Voice in the Program

There are written and oral course evaluations during and after each course, as well as regular dialogue with student representatives. The program is appreciated by students due to its clear structure with lectures and follow-up seminars, the engagement shown by the mathematics teacher educators, and the relevance of the content taught with respect to both content knowledge and PCK. Students who had a negative attitude towards mathematics previously declare developing a more positive attitude towards the subject. They learn useful methods for teaching mathematics and enjoy teaching it—a subject with which they previously struggled! In particular, the bridge between content matter and PCK is highly appreciated.

Challenges

One major challenge is the consequence of the academization of teacher education, including the shift from a college tradition to an academic tradition (Sjöberg, 2018). Because Sweden is a small country with a limited research community in mathematics education and lacks research journals that publish in Swedish, academics from other countries have been brought to strengthen the connections between research and teacher education. Also, research articles written in English are utilized as part of the course materials. Naturally, as a result, language issues arise, where some lecturers struggle to explain mathematics and mathematics education terms in Swedish.

EDUCATION OF UPPER PRIMARY MATHEMATICS TEACHERS

In the following section, the program implementation at Stockholm University will be presented as one case. As there are many similarities in both content and structure among Swedish universities, this is not an exceptional case. It simply allows us to paint a clear picture of the program.

Over the four years of the program, students study one semester (30 ECTS) of mathematics/mathematics education, divided into three courses. The courses integrate the domains of mathematical knowledge for teaching, and the design of the courses is guided by the model from Ball et al. (2008) and Hill and Ball (2009). Consequently, elements of *subject matter knowledge* (SMK) and PCK are intertwined to provide a structure for the students to think about mathematics teaching. Courses are taught by both mathematics lecturers and lecturers specialized in mathematics education. The aim for coherence and progression between the courses and the striving for students to deepen their *knowledge of content and students* (KCS) and *knowledge of content and teaching* (KCT) have inspired us to structure the two first courses similarly. The mathematical content is introduced by refreshing the students' own arithmetical and geometrical skills. This boost of the students' own mathematical skills is similarly structured in the two first courses; it is later connected to the students' future profession as a mathematics teacher through activities and seminars addressing the aspects of teaching and learning mathematics that are most relevant for the practicum. The first course focuses mainly on numeracy/number sense and arithmetic relevant for grades 4–6, though it touches upon number sense for younger children and looks a bit further towards grades 7–9. The geometry course has the same approach, but we deepen elements of PCK for grades 4–6 in activities and literature seminars. The third course is the largest and contains algebra, functions, statistics, and probability. There is a strong focus on the three PCK elements (KCS, KCT, and knowledge of curriculum) throughout, although algebra, statistics, and probability are challenging areas of mathematics for many students, which require separate attention to revisiting content.

The PCK component is designed so the students gradually get greater responsibility for teaching and planning how to teach. One example of this is the strand of "micro teaching" over the three courses that allows students: (1) to explain specific content connected to numbers or arithmetic to peer students (i.e., concepts as well as theorems, proofs, and course routines); (2) to plan an activity for learning geometry concepts and present it to peers for critique; and (3) in the last course plan a mathematical path outside the university or in a museum (extramural learning). The latter task is organized so that school classes come to walk this path, engaging in activities connected with different mathematical content. The students hence realize and take responsibility for the activities with the children. The following day, together with a lecturer, the students reflect on mathematics education questions, challenges, and opportunities for learning.

The lecturers always aim to make both content and pedagogy visible for the becoming teachers, so that they are learning about learning mathematics from both the teachers' and the learners' perspectives at the same time. This is accomplished through activities, use of tools, and strategies for learning mathematics that can facilitate both their own and future learners' mathematics acquisition as well as their views on and attitudes toward mathematical tasks. Information and Communication Technology (ICT), manipulatives, and other visual means are used in all three courses, not as something extraordinary, but as artefacts for mathematics teaching and learning generally. In the last course, the students' task is to make a film with iPads wherein they must introduce one mathematical concept (e.g., probability).

The students also work with programming in the third course. They learn how to do basic visual programming in computer programs for young children. They have to present the resulting programs to the class and engage mathematics education questions, what they want learners to learn, why, and how. The programs are sent to a couple of classes in schools for testing and the learners give the students feedback on their work. During the third course, students develop planning for an entire content area, motivating their choices with reference to the course readings. Finally, they create an evaluative instrument and an instrument for learners' self-

---

**Vignette 2.6 The Role of the Contexts
(Constituting the Ecology) of a Teacher's Practice**

The Contexts of Mary's Mathematics Teaching

It is Friday morning and the grade 5 teacher Mary enters the classroom filled with excitement. Friday morning means problem solving, and this is the main reason for the excitement. Mary loves when the learners engage with problems, sometimes a bit out of their reach, collaborate, think outside the box, and explore different solutions. The best part of the lesson is at the end when the learners discuss their solutions. With a small sigh, Mary thinks about the teacher meeting last Tuesday. She and her colleagues who also teach grade five discussed their teaching, but Mary was not content. She asked for more discussion about how to teach but that never happens, there is always something immediate that needs to be solved first. Mary longs for deeper discussions and a focus on teaching, not what pages in a book to cover or what to include on a test.

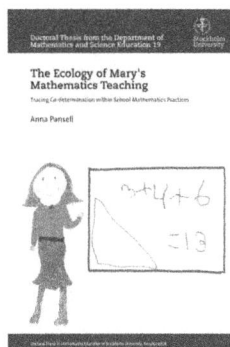

This example is based on observations and interviews with one mathematics teacher in grade five from a single case study (Pansell, 2018). The study found a focus on how to teach, not on arguments and theoretical grounds for why mathematics should be taught in certain ways. This was not only visible in teachers' practices; the lack of theoretical arguments for mathematics teaching was also visible in mathematics textbooks, teacher guides, and the national curriculum. For Mary this means that there is no forum where she can develop her teaching of problem-solving. She is left to develop her teaching by herself with little influence from research-based literature—an influence she would have welcomed.

---

assessment. The intention is for the students to use these tasks as opportunities for preparing for their last practicum, and many students do so.

Some of the students have participated in the TRACE project,[17] and observations of their classroom practice indicate a strong focus on the learners and their learning, as well as on varied teaching and giving space for the many creative and non-conventional ways in which children approach problems (see also Vignette 2.6). This is in line with what is encouraged in their education and it suggests to us that we are on the right path; however, there is still room for improvement.

LOWER SECONDARY MATHEMATICS TEACHER EDUCATION

Lower secondary mathematics teacher education in Sweden has a history of being caught between the demands of compulsory school, with its broader perspec-

---

[17] https://www.mnd.su.se/english/research/mathematics-education/research-projects/trace

tive on education for all, and upper secondary education, with its specific demands for in-depth subject teaching. This has resulted in shuttling between two approaches. From the 1960s to late 1980s, the education of teachers for lower and upper secondary levels was mainly conducted as a subject-teacher education, where students first obtained a Bachelor's or Master's in two disciplines and then completed a year of practice teaching and studies in pedagogy. In 1988, the subject-teacher program was changed, and the lower secondary teacher education program was connected more firmly to the needs of compulsory school. Hence, the lower secondary mathematics teachers were prepared for teaching grades 4–9 in three to five subjects. Mathematics was combined with science and technology.

However, in the present teacher education program (since 2011), the pendulum has more or less returned to the subject-teacher tradition, and lower and upper secondary teacher education are once again joined. The students study for between four and five and a half years to the level of Bachelor's or Master's, preparing them to teach in two subjects. The length of the program depends on the subject specializations. For instance, students who want to combine mathematics with either Swedish or social studies will need to spend 5½ years. The students can choose to study mathematics as their major (90 ECTS) or their minor (60 ECTS). It is also possible for students who already have a Bachelor's or Master's in mathematics (as a minor or a major) to study for 1–1.5 years to obtain a teaching degree for secondary school.

The Case of One Program—Content

At Karlstad University and its predecessors, teacher education has been offered since the 1840s, and secondary mathematics teacher education since the late 1960s. Lower and upper secondary teacher education have generally been offered together. During some periods, the former was conducted as a shorter version of the upper secondary teacher program, and at other times it has been a program on its own.

In the short program, students study general pedagogy, subject education, and complete a practicum, but are assumed to have acquired content knowledge in their previous degree. One of the short programs has, for some time, been integrated into the practices of local schools. However, this section focuses on the content of the longer program.

The aim of the program is to develop a knowledge base for mathematics teaching based on knowledge of the needs in compulsory school (related to its organisation, policies, and curriculum, as well as the mathematical trajectory in compulsory school). Hence, both content knowledge and pedagogical knowledge are important objects of study. The mathematics content/mathematics education courses are, to a large extent, held jointly with upper secondary education student teachers, but separate from engineering programs, scientific programs, and so on. That is, the mathematics courses are directed towards "mathematics for teachers."

The mathematical content in the first semester focuses on problem solving, introductory algebra, and analysis. Mathematics education issues concerning policies and curricular materials for compulsory school, learners' development of number sense, problem solving strategies, and mathematical thinking are strongly linked to the mathematical content. Teaching proofs and proving are addressed, as well as the history of mathematics. The students work with dynamic software to visualize and design inquiry-based activities in "micro lessons." The students work on developing both their own mathematical competency and their PCK simultaneously. The assessments are individual written exams, oral presentations, group exercises where they develop mathematics lessons using dynamic software, and writing reports individually or in groups.

In the second semester, the mathematical content focuses on linear algebra, probability and statistics, modeling and differential equations, and geometry. Digital tools for teaching and learning are used, and mathematics education aspects are studied. Also, in this semester assessment is individual or in groups, oral or written, or digitally developed and presented. Micro lessons, lesson plans, and reports are presented and scrutinized in seminars.

The third semester contains a course in mathematics focusing on discrete mathematics and algebraic structures. This course is followed by a course in general pedagogy focusing on theories of formative and summative assessment, and then a mathematics education course focusing on mathematical language and the relation between language and mathematical learning, individualization of mathematics teaching (such as special needs and gifted learners), curricular material from an international perspective, and a mandatory practicum. Assessments are both individual and in groups.

In the rest of the program, the students study their second subject, courses in general pedagogy, complete a practicum, and complete independent research projects—one on the Bachelor's level (15 ECTS) and one on the Master's level (15 ECTS). If the student chose mathematics as a major teaching subject, the Master's thesis will be on mathematics or mathematics education.

Pedagogical Approach Used in the Program

The pedagogical approach from the lecturers teaching mathematics or mathematics education takes its departure from mathematics as a discipline and its transformation into a school subject. Developing the students' mathematical knowledge is one cornerstone, but the lectures also pay attention to the development of pedagogical knowledge for teaching. In order to conduct well-designed instruction, the student teachers need to develop a competency to transform learning goals into activities that make it possible for learners to learn. The aim of the program is to fulfil the needs of a lower secondary teacher regarding all aspects of mathematical knowledge for teaching (see e.g., Ball et al., 2008). Hence, different teaching modes include lectures (online, as well as *in situ*), exercises, and investigative activities with hands-on materials or digital tools. The choice and implementation of modes are continuously developed and refined. The lecturers managing the program have different competences: mathematicians, mathematics educators, and school-based mentors. This means that the competences cover content knowledge; curricular knowledge; knowledge of learners and their characteristics; knowledge of educational contexts, educational ends, and the purposes and values of education; and PCK. Consequently, the mathematics/mathematics education courses in the program are able to cover all these aspects of mathematics teachers' knowledge base.

Because many of the teacher educators at Karlstad University are active researchers, specifically focusing on digital tools for developing teaching of mathematical competencies, their research influences both the content and organization of the courses. The literature used in the courses includes mathematics textbooks in English and Swedish, papers from peer reviewed international journals in mathematics education, literature from general pedagogics, literature from Swedish mathematics education research, as well as guidelines, curricular materials, steering documents, and so on from the Ministry of Education. In the courses, the students also review textbooks for compulsory school, investigate on-line teaching material, and explore and develop digital tools (e.g., apps and other software) for mathematics teaching. The recent demands for digitizing school, as well as the cross-curricular inclusion of programming in compulsory school, have expanded these kinds of activities in teacher education as a whole, and specifically in the mathematics/mathematics education courses.

EDUCATION OF UPPER SECONDARY MATHEMATICS TEACHERS

This section outlines the case of the program as it is implemented at Umeå University. It addresses the overall structure of the program, the courses in the program as well as the teaching methods, and the use of research in the program, before discussing a few current issues.

The Structure of the Program

In Umeå, all upper secondary teachers study (at least) two main subjects. Mathematics can be combined with all subjects (e.g., English, Physics, Chemistry, Geography, or History) except for Swedish. Of the five-year program, two years are courses in mathematics and mathematics education (provided that the final Master's thesis is written in the area of mathematics education), one and a half years are courses in the second subject, half a year is practicum experiences, and one year of courses are in general educational sciences. The layout of courses is shown in Figure 2.1. All students write a Master's thesis at the end of their preservice education, but they can choose whether to write this in mathematics education or in their second subject. In the latter case, the second subject will comprise 120 ECTS and mathematics will comprise 90 ECTS. This is a difference

Semester	1	2	3	4	5	6	7	8	9	10
Educational Science	█				█					
Maths and Math. Ed.		█	█	█					█	**
Second Subject						█	█	█		
Practicum	█				█			█		

FIGURE 2.1.   The Layout of Courses Over the Mathematics Teacher Program With Mathematics as the Student's First Subject.
Note: ** indicates the independent research project.

between universities, where some universities have chosen to include two smaller theses (15 ECTS), one in each subject.

Courses

The courses in mathematics and mathematics education include courses in pure mathematics, such as linear algebra, calculus, and discrete mathematics. The preservice teachers take a few of these courses together with engineering students, but most of these mathematics content courses are designed especially for secondary (lower and upper) preservice teachers and focus more on theory (mathematical and mathematics education) and a deep conceptual understanding of mathematics. Courses designed especially for these students include mathematical methods, algebra, precalculus and calculus, differential equations and multivariable calculus, and mathematics education. The courses in mathematics education tie together the mathematics courses and practicum with a focus on the mathematical competencies described in the curriculum for upper secondary school (Swedish National Agency for Education, 2011), mathematics education, and practical elements. In contrast to the mathematics courses that focus on higher level mathematics, the mathematics education courses include both a basic mathematical content focus (e.g., number sense, fractions, (pre-)algebra, geometry, or calculus) and a focus on mathematical competencies (e.g., concepts, procedures, reasoning, modeling, and problem-solving). The students discuss how learners' conceptions can hinder their learning, and how mathematical competencies can be interpreted and implemented in school. Frameworks that focus on mathematical competencies (e.g., National Governors Association Center for Best Practices & Council of Chief State School Officers, 2010; Niss & Jensen, 2002) as well as research articles and books that provide a basic foundation of the mathematics education field are discussed (e.g., Boaler, 1998; Kieran, 1981; Schoenfeld, 1992). Lectures are mixed with practical exercises that include work with common learner conceptions, different teaching methods (see the following section), lesson planning, or solving tasks that give the students an idea of the difficulties learners encounter. For example, the students get to make calculations with new numbers (zero, do, re, mi, fa, so) in base six to get an idea of the problems learners can have with simple calculations if their number sense is underdeveloped.

Lesson planning, concretizing aims in the curriculum, and connecting this to a classroom activity are important parts of being a teacher. The students plan for a single lesson during course one, a four-lesson segment in course two, and finally a whole section (some four to five weeks) during the last practicum.

Mathematics education is the focus of the final Master's thesis, during which students have to prepare, carry out, and report on a small research study, as well as defend their thesis and act as a critical reader of a peer's thesis. See also the section on the independent research project.

Teaching Methods

Within the program, students meet a variety of teaching methods. In the larger student groups (i.e., when taking courses together with engineering students), classical lectures with successive lessons in smaller groups are most common. It is also possible for students to meet in small groups outside the classroom, where two tutors are always available if support is needed. In the special teacher courses, there is greater variety in the teaching methods that the students meet. The students work in groups, discuss situations that they will encounter as teachers, have seminars where they discuss research articles, or solve mathematical problems. There are elements of outdoor mathematics, practical mathematics labs (both hands-on and in computer labs), and practice presentations.

All this is done partly to prepare the students for the upcoming practicum, where they meet learners and will be responsible for some part of their mathematics education, and partly to show the students that mathematics can be taught in different ways. Most students have only met traditional teaching methods with lectures and individual work with textbooks through their own schooling (Boesen et al., 2014; Mullis et al., 2012), and students have repeatedly described in course evaluations that encountering different methods is an eye-opener.

Use of Research

Most lecturers teaching in the program have a Ph.D., and are active researchers in mathematics or mathematics education. Umeå University has one of the largest Mathematics Education Research Centres in Scandinavia, and the proximity to this active research environment provides students with opportunities to participate in research seminars, research studies, and to have active researchers as supervisors for their final Master's thesis. The literature in the mathematics and mathematics education courses comprises textbooks in Swedish and English, articles from international research journals or conferences, and papers from the governmental inservice mathematics education program, as well as curricula and guides provided by the Swedish National Agency of Education. Although there is a close connection to the Umeå mathematics education research group, which somewhat can influence the academic language and scientific methods of assignments and Master's theses, practical experience is also considered valuable for the preservice teachers. The teachers of the mathematics education courses all have (besides their Ph.D.) a teacher degree and experience from teaching in upper secondary schools.

Issues

Some students in the program have Swedish as an additional language (i.e., not as their first language). Mostly this is not an issue, but at times it provides challenges to being or becoming a teacher. Teachers have to be able to communicate not only with learners and their parents, but also with colleagues. As the preservice teachers are expected to teach in Sweden, learning to communicate in Swedish is crucial. A few assignments can be written in English,[18] but practicum is conducted in Swedish. Umeå University provides support for students who struggle with Swedish, whether it be writing academic texts, due to dyslexia or struggling with Swedish as an additional language, but it is imperative that the students are motivated to seek help when needed.

Upper secondary teachers are also allowed to teach in lower secondary school. Because the preservice upper secondary teachers take most courses together with preservice lower secondary teachers, they get a theoretical idea of what teaching in lower secondary entails. Previously, it was not compulsory to do any of the three practicums in lower secondary school, but this was changed because learners in their early teens behave differently from upper secondary learners, and because the subject of mathematics is more abstract in upper secondary school. To prepare the students for this, they are provided with activities where they consider the curriculum for lower secondary mathematics, and where textbook analysis and practice planning also includes the lower secondary level. Lately, it was decided that all upper secondary student teachers should do the second practicum (4 weeks) in lower secondary school to ensure that they have some hands-on experience with lower secondary mathematics and learners.

At present, Sweden is in great need of mathematics, science, and technology teachers. A variety of routes to becoming a mathematics teacher are therefore being developed, such as reskilling of mathematics, science, or engineering graduates wanting to change careers or special programs for recent immigrants. Vignette 2.7 describes a different initiative, a unique program that combines engineering and teacher education.

---

[18] Discussions at the moment could result in fewer possibilities for students to write assignments in another language than Swedish, except for preservice language teachers (currently French, German, Spanish, and English).

> ### Vignette 2.7. The Structure of a "Double Degree" Program Combining Engineering and Teacher Education
>
> Since 2002 there exists a five-year, double degree program called *Master of Science in Engineering and Education*. KTH Royal Institute of Technology in Stockholm offers the program in cooperation with Stockholm University. After five years of study, which is the stipulated time for both the teacher program and the engineering programs in general, the graduates of the double degree program become engineers *and* upper secondary teachers of mathematics (and technology, physics, or chemistry). This is made possible by considering the teaching profession as a specialization within engineering and treating engineering knowledge and skills as subject knowledge in Mathematics, Science, and Technology teacher education.
>
> Advantages with this approach are that students have a broad choice of potential career. They can change career pathways, and some individuals become teachers who might not have considered this profession otherwise. Disadvantages are the very packed program and the challenges students face in bridging two "cultures."
>
> The impact of this program on upper secondary schools in the Stockholm region has been significant. Over recent years, there have been around 500 applicants and 60 students admitted to the program yearly. Surveys sent out to former students within a year of graduation show that about 30% of the graduates work as mathematics teachers.
>
> This particular engineering and teacher education program is structured throughout with the parallel teaching of mathematics courses (oriented for both teaching and engineering), general pedagogical courses, mathematics education courses, and practicum periods.
>
> In the first year of the program, students take courses on teaching as a profession, programming (especially *Matlab)*, the conceptual world of mathematics and mathematical reasoning and learning, discrete mathematics, and mathematics education. This is followed by a practicum.
>
> The second year includes a course on social relations and educational leadership and an extensive number of mathematical courses (e.g., *Linear Algebra, Analysis in One Variable, Multivariable Calculus, and Vector Analysis*).
>
> In the third year, students take a course on organisational analysis and professional roles, and a lot more mathematics (*Numerical Methods, Differential Equations and Transformations, Probability Theory,* and *Statistics*).
>
> The fourth year gives students even more mathematics through an advanced course. But the program also makes it possible for students to learn about project management and business development, teaching and sustainable development, learning and assessment in mathematics, other mathematical education, and finally a second school placement. During the fourth year, the students have to choose two elective mathematics courses—for example, *Complex Analysis, Foundations of Analysis, Financial Mathematics,* or *Mathematical Systems Theory.*

> **Vignette 2.7.   Continued**
>
> In the fifth and final year, the program includes courses on theory of knowledge/philosophy of science and research methodology for teachers, the history of mathematics, curriculum theory, overviews of mathematics education research, and special education. The last term contains the final Degree Project, which is required to combine Technology/Mathematics and Education.
>
> This specific program can be understood as an innovative mathematics teacher education that gives both competence to work as an engineer with a pedagogical profile and as a teacher with an engineering perspective. In that respect, the program combines pedagogical competence, communication, and learning with the skills of an engineer, dealing with new knowledge and problem solving. Those perspectives also reflect the nature of mathematics teaching and learning within the program. Not least, this program presents a specific vision of mathematics teaching and learning, and of the mathematics teaching profession.

DISCUSSION

The descriptions of program implementation at different institutions show how the focus varies; foregrounding "teaching method" more or less, linking mathematics and mathematics education more or less strongly, and so forth. This also varies with the level of schooling for which the student teachers are being prepared—which is common internationally and hence not surprising. However, it is difficult to draw out further comparisons, as institutions have also chosen to accentuate different dimensions in their case descriptions.

Looking across narratives, descriptions of the history of and current national guidelines for mathematics teacher education, and the cases from the institutions, a development towards increased academization is clear. This is a key aspect of the dilemma of both increasing the number of graduates from the programs and ensuring a high-quality research-based preparation for the profession. This dilemma is well described internationally. In the USA, it has led to alternative certification routes (Aragon, 2016), but not so in Sweden where the pressure is rather to increase the academic scholarship component of teacher education.

Several of the institutional cases address the connections between mathematics content and mathematics education, and between campus courses and practicum experiences, as well as the difficulties in ensuring a two-way relationship. A review of the international literature on interventions aimed at closing the so-called theory-practice gap in mathematics teacher education showed that "theory" was often perceived to be what is taught on campus, and guidelines for students' and teachers' practices (Taub, 2019). This view is not reflected in the descriptions in this chapter, which indicate a view of the theory-practice connection as having multiple dimensions and working in both directions. At the same time, the national regulations and guidelines to some extent pull in opposite directions. On the one hand, they point to the need for integration of theory and practice; on the other hand, they solidify the practicum as a separate component, detached from mathematics education theory and research. The pressure on institutions to overcome this in their implementation of programs is evident across several of the sections in this chapter. In the discussion of reflections on practicum experiences, it is clear how the lack of explicitness around the need to link reflections to theory/research generates problems. Conversely, some descriptions in this chapter of the different program implementations share positive results in this respect.

The division of authority between the nation-state and municipality (or private boards of education in schools that are run by businesses or organizations) causes a tension in the debate about schools. School politics are frequently raised in election debates, and results of international tests such as TIMSS and PISA are used in the argument for school reform. The latest reform demanding an implementation of computational thinking and programming illustrates the process of change in Swedish schools, and the external, overwhelmingly political forces at play. The revised school curriculum was published in 2017 and was to be implemented within a year. Without any previous inservice training or adjusted university programs, mathematics teachers on all school levels, from kindergarten to teacher education, were expected to include this new content in their teaching. Executing the reform is a municipal responsibility, but ultimately the short time span from signing into law to implementation has sent both teachers and teacher educators scuttling.

We have highlighted issues around diversity, including challenges related to language. That roughly a quarter of learners are entitled to study a mother tongue different from Swedish can be taken as an indication of the cultural and linguistic diversity within which schools (and teacher education) must operate. Unlike countries with a longer history of a diverse population, this requires urgent adjustments in teaching and teacher education. Learners' experiences of exclusion must be considered (see Svensson Källberg, 2018), as must issues of access and fairness (see Hinnerich et al., 2015; Petersson, 2017). We believe there is a need in teacher education for addressing inclusion, as well as for making teacher education itself inclusive. Besides language assistance, this may require making expectations more explicit (see Österling, 2021). Currently, such considerations and initiatives are up to the individual institutions.

Overall, this chapter—not unexpectedly—points to mathematics teacher education being reliant on the socio-political, cultural, and administrative systems within which it operates, including trying to realize its required—and perhaps desired—commitment to research-based practices and a comprehensive notion of mathematics as a discipline. One aspect we have not addressed in this chapter, but which is of utmost importance to how we discuss mathematics teacher education, is the notion of mathematics within both national curricula and institutional program implementation. It is possible to read between the lines that there are differences, but also that a cognitive perspective on mathematics learning appears to dominate. It is going to be interesting to follow the research and teacher education practices as a new generation of researchers may bring in more discursive or socio-political perspectives (see Christiansen & Skog, in preparation; Helenius et al., 2018). The EU-guidelines, the national political situation, and the opposite pulls of visions and regulations constitute the crosscurrents within which Swedish teacher education operates, but simultaneously generate a crevice of opportunities for critical visions and forward-thinking program development.

ACKNOWLEDGMENTS

This chapter draws on research conducted in the TRACE project (Tracing Mathematics Teacher Education in Practice, see footnote 17) with funding from the National Research Council, project 2017-03614.

REFERENCES

Adler, J., & Davis, Z. (2006). Opening another black box: Researching mathematics for teaching in mathematics teacher education. *Journal for Research in Mathematics Education, 37*(4), 270–296.

Aragon, S. (2016). *Teacher shortages: What we know*. Teacher Shortage Series. Education Commission of the States.

Åstrand, B. (2016). From citizens into consumers: The transformation of democratic ideals into school markets in Sweden. In F. Adamson, B. Åstrand, & L. Darling-Hammond (Eds.), *Global education reform: How privatization and public investment influence education outcomes* (pp. 79–115). Routledge.

Ball, D. L., Thames, M. H., & Phelps, G. (2008). Content knowledge for teaching: What makes it special? *Journal of Teacher Education, 59*(5), 389–407.

Beach, D., Bagley, C., Eriksson, A., & Player-Koro, C. (2014). Changing teacher education in Sweden: Using meta-ethnographic analysis to understand and describe policy making and educational changes. *Teaching and Teacher Education, 44*, 160–167.

Bejerot, E., Hasselbladh, H., Forsberg, T., Parding, K., Sehlstedt, T., & Westerlund, J. (2018). Förberedd för läraryrket? Lärare under 40 år av reformer [Prepared for the teacher profession? Teachers during 40 years of reforms]. *Arbetsmarknad & Arbetsliv* [Labour Market and Working Life], *24*(1–2), 7–26.

Bennett, J., & Tayler, C. P. (2006). *Starting strong II. Early childhood education and care*. OECD.

Bentley, P-O. (2003). *Mathematics teachers and their teaching—A survey study*. Doctoral Thesis from University of Gothenburg, Gothenburg Studies in Educational Sciences.

Biesta, G. (2013). Knowledge, judgement and the curriculum: On the past, present and future of the idea of the practical. *Journal of Curriculum Studies, 45*(5), 684–696.

Boaler, J. (1998). Open and closed mathematics: Student experiences and understandings. *Journal for Research in Mathematics Education, 29*(1), 41–62.

Boesen, J., Helenius, O., Bergqvist, E., Bergqvist, T., Lithner, J., Palm, T., & Palmberg, B. (2014). Developing mathematical competence: From the intended to the enacted curriculum. *The Journal of Mathematical Behavior, 33*, 72–87.

Calander, F. (2008). Lärarutbildning som symboliskt kapital i ett förändrat utbildningslandskap [Teacher education as symbolic capital in a changed educational landscape]. In M. Carlsson & A. Rabo (Eds.), *Uppdrag mångfald: Lärarutbildning i omvandling* [Teacher education under transformation] (pp. 259–293). Boréa.

Christiansen, I., Bertram, C., & Mukeredzi, T. (2018). Contexts and concepts: Analysing learning tasks in a foundation phase teacher education programme in South Africa. *Asia-Pacific Journal of Teacher Education, 46*(5), 511–526.

Christiansen, I. M., Österling, L., & Skog, K. (2019). The desired teacher in six observation protocols. *Research Papers in Education*, DOI: 10.1080/02671522.2019.1678064

Christiansen, I. M. & Skog, K. (in preparation). Ett snapshot av svensk matematikdidaktisk forskning. [A snapshot of Swedish mathematics education research]. In P. Valero, L. Björklund-Boistrup, I. M. Christiansen, & E. Norén (Eds.), *Matematikundervisning i samhället: Sociopolitiska utmaningar* [Mathematics instruction in society: Sociopolitical challenges].

Dahlstedt, M., Harling, M., Trumberg, A., Urban, S., & Vesterberg, V. (2019). *Fostran till valfrihet: Skolvalet, jämlikheten och framtiden* [Raising to freedom of choice: The school choice, the equity, and the future]. Liber.

Dysthe, O., & Engelsen, K. (2004). Portfolios and assessment in teacher education in Norway: A theory-based discussion of different models in two sites. *Assessment & Evaluation in Higher Education*, *29*(2), 239–258.

Eidevald, C., & Engdahl, I. (2018). *Utbildning och undervisning i förskolan: Omsorgsfullt och lekfullt stöd för lärande och utveckling* [Education and teaching in preschool: Caring and playful support for learning and development] (1st ed.). Liber.

Ellegård, K., & Vrotsou, K. (2013). *En tidsgeografisk studie av strukturen i lärares vardag* [A time-geographical study of the structure of teachers' days]. Skolverket.

Emsheimer, P., & Silva, N. L. D. (2011). Preservice teachers' reflections on practice in relation to theories. In M. Mattsson, T. V. Eilertsen, & D. Rorrison (Eds.), *A practicum turn in teacher education* (pp. 147–167). Sense. https://doi.org/10.1007/978-94-6091-711-0_8

Forssell, A., & Ivarsson Westerberg, A. (2014). *Administrationssamhället* [The administration society]. Studentlitteratur.

Fransson, O. (2009). Epistemisk förskjutning och autonomi [Epistemic shift and autonomy]. In O. Fransson & K. Jonnergård (Eds.), *Kunskapsbehov och nya kompetenser: Professioner i förhandling* [Knowledge needs and new competencies: Professions under negotiation] (pp. 21–44). Santérus.

Gabrielsson, A. (2005). *Utvärdering av den nya lärarutbildningen vid svenska universitet och högskolor* [Evaluation of the new teacher education at Swedish universities and tertiary colleges]. Högskoleverket.

Gawell, E. (2017). "Men det stämmer inte riktigt med hur svenskarna räknar": Uppfattningar om skolmatematik och matematikundervisning hos gymnasieelever med utländsk grundskoleutbildning ["But that's not really how the Swedes do math": Immigrant high school learners' conceptions of school mathematics and math education]. Independent thesis Advanced level (professional degree). Faculty of Science. Stockholm University.

Helenius, O., Kilhamn, C., & Nyström, P. (2018). Mathematics education research in Sweden: National presentation at PME 42. In E. Bergqvist, M. Österholm, C. Granberg, & L. Sumpter (Eds.), *Proceedings of the 42nd Conference of the International Group for the Psychology of Mathematics Education* (Vol. 1, pp. 273–297). Sweden.

Hill, H., & Ball, D. L. (2009). The curious—and crucial—case of mathematical knowledge for teaching. *Phi Delta Kappan*, *91*(2), 68–71. https://doi.org/10.1177/0031721709100215

Hinnerich, B. T., Höglin, E., & Johannesson, M. (2015). Discrimination against students with foreign backgrounds: Evidence from grading in Swedish public high schools. *Education Economics*, *23*(6), 660–676.

Imsen, G., Blossing, U., & Moos, L. (2017). Reshaping the Nordic education model in an era of efficiency. Changes in the comprehensive school project in Denmark, Norway, and Sweden since the millennium. *Scandinavian Journal of Educational Research, 61*(5), 568–583.

Jons, L. (2019). The supportive character of teacher education triadic conferences: Detailing the formative feedback conveyed. *European Journal of Teacher Education, 42*(1), 116–130.

Karlsudd, P. (2018). Att problematisera "problemet": Bedömning och utveckling av problemformuleringar i lärarutbildningens självständiga arbeten [To problematize "the problem": Assessment and development of problem statements in the independent research project in teacher education]. *Nordic Journal of Vocational Education and Training, 8*(1), 1–22.

Kieran, C. (1981). Concepts associated with the equality symbol. *Educational Studies in Mathematics, 12,* 317–326.

Kilborn, W. (1992). *Didaktisk ämnesteori i matematik 1–3* [Didactical/educational subject theory of mathematics 1–3]. Utbildningsförlaget.

Kilhamn, C. (2011). *Making sense of negative numbers.* Doctoral Thesis from University of Gothenburg/Gothenburg Studies in Educational Sciences.

Kilhamn, C., Nyman, R., Knutsson, L., Holmberg, B., Frisk, S., Skodras, C., & Gallos Cronberg, F. (2019). *Mathematical conversations—Ways to pupils' learning.* Liber.

Kirschner, P. A., Sweller, J., & Clark, R. E. (2006). Why minimal guidance during instruction does not work: An analysis of the failure of constructivist, discovery, problem-based, experiential, and inquiry-based teaching. *Educational Psychologist, 41*(2), 75–86.

Korthagen, F., & Vasalos, A. (2005). Levels in reflection: Core reflection as a means to enhance professional growth. *Teachers and Teaching, 11*(1), 47–71.

Löwing, M. (2004). *Matematikundervisningens konkreta gestaltning. En studie av kommunikationen lärare-elev och matematiklektionens didaktiska ramar.* [The concrete formation of mathematics teaching: A study of communication between teachers and pupils and the educational framework of mathematics classrooms.] Doctoral Thesis from University of Gothenburg, Gothenburg Studies in Educational Sciences.

Maaranen, K. (2010). Teacher students' MA Theses—A gateway to analytic thinking about teaching? A case study of Finnish primary school teachers. *Scandinavian Journal of Educational Research, 54*(5), 487–500.

Meeus, W., Van Looy, L., & Libotton, A. (2004). The bachelor's thesis in teacher education. *European Journal of Teacher Education, 27*(3), 299–321.

Melén, B., & Wedman, I. (n.d.). *Betyg i svensk grundskola- och gymnasieutbildning* [Grades in Swedish primary and secondary education]. https://www.ne.se/uppslagsverk/encyklopedi/l%C3%A5ng/betyg

Mickwitz, L. (2015). *En reformerad lärare: Konstruktionen av en professionell och betygssättande lärare i skolpolitik och skolpraktik* [A reformed teacher: The construction of a professional and grading teacher in school politics and school practice]. Doctoral Dissertation, Stockholm University, Department of Education.

Mullis, I. V., Martin, M. O., Foy, P., & Arora, A. (2012). *TIMSS 2011 international results in mathematics.* International Association for the Evaluation of Educational Achievement.

National Agency for Education. (2018). *Curriculum for the preschool. Lpfö 18.* National Agency for Education.

National Agency for Education. (2019). *Så här fungerar förskolan* [This is how preschool works]. https://www.skolverket.se/for-dig-som-ar.../elev-eller-foralder/skolans-organisation/sa-har-fungerar-forskolan

National Governors Association Center for Best Practices & Council of Chief State School Officers. (2010). *Common core state standards for mathematics*. Author.

Niemelä, P. S., & Helevirta, M. (2017). K–12 curriculum research: The chicken and the egg of math-aided ICT teaching. *International Journal of Modern Education & Computer Science, 9*(1), 1–14.

Niss, M., & Jensen, T. H. (2002). *Kompetencer og matematiklæring: Idéer og inspiration til udvikling af matematikundervisning i Danmark* [Competences and mathematics learning: Ideas and inspiration for the development of mathematics education in Denmark] (Vol. 18). Undervisningsministeriet.

Organisation for Economic Co-operation and Development (OECD). (2012). *Education at a glance 2012: OECD indicators. Country notes: Sweden*. Author. https://www.oecd.org/sweden/EAG2012%20-%20Country%20note%20-%20Sweden5.pdf

Österling, L. (in preparation). Beställningen av den dugliga läraren [The requisition of the capable teacher]. In P. Valero, L. Björklund-Boistrup, I. M. Christiansen, & E. Norén (Eds.), *Matematikundervisning i samhället: Sociopolitiska utmaningar* [Mathematics instruction in society: Sociopolitical challenges].

Österling, L. (2020). *Confessions of mathematics student teachers*. Unpublished manuscript. Stockholm University.

Österling, L. (2021). inVisible theory in pre-service mathematics teachers' practicum tasks. *Scandinavian Journal of Educational Research, 65*. DOI:10.1080/00313831.2021.1897874

Palmér, H., & Björklund, C. (2016). Different perspectives on possible—Desirable—Plausible mathematics learning in preschool. *Nordic Studies in Mathematics Education, 21*(4), 177–191.

Pansell, A. (2018). *The ecology of Mary's mathematics teaching: Tracing co-determination within school mathematics practices*. Doctoral dissertation, Department of Mathematics and Science Education, Stockholm University.

Petersson, J. (2017). *Mathematics achievement of early and newly immigrated students in different topics of mathematics*. Doctoral dissertation, Department of Mathematics and Science Education, Stockholm University.

Pramling, N., Wallerstedt, C., Lagerlöf, P., Björklund, C., Kultti, A., Palmér, H., Magnusson, M., Thulin, S., Jonsson, A., & Pramling Samuelsson, I. (2019). *Play-responsive teaching in early childhood education*. Springer.

Råde, A. (2014). Ett examensarbete för både yrke och akademi—En utmaning för lärarutbildningen [A degree project for both profession and academy—A challenge for teacher education]. *Högre Utbildning* [Higher Education], *4*(1), 19–34.

Reikerås, E., Løge, I. K., & Knivsberg, A-M. (2012). The mathematical competencies of toddlers expressed in their play and daily life activities in Norwegian kindergartens. *International Journal of Early Childhood, 44*(1), 91–114.

Reis-Jorge, J. (2007). Teacher's conceptions of teacher-research and self-perceptions as enquiring practitioner: A longitudinal case study. *Teaching and Teacher Education, 23*(4), 402–417.

Rønning, F. (2019). Didactics of mathematics as a research field in Scandinavia. In W. Blum, M. Artigue, M. A. Mariotti, R. Sträßer, & M. Van den Heuvel-Panhuizen (Eds.), *European traditions in didactics of mathematics* (pp. 153–185). Springer.

Rusznyak, L. (2012). Summative assessment of student teaching: A proposed approach for quantifying practice. *Journal of Education*, *56*, 91–115.

Rusznyak, L., & Bertram, C. (2015). Knowledge and judgement for assessing student teaching: A cross-institutional analysis of teaching practicum assessment instruments. *Journal of Education*, *60*, 31–61.

Schoenfeld, A. H. (1992). Learning to think mathematically: Problem solving, metacognition, and sense making in mathematics. In D. A. Grouws (Ed.), *Handbook of research on mathematics teaching and learning* (pp. 334–370). Macmillan.

Schön, D. A. (1987). *Educating the reflective practitioner: Toward a new design for teaching and learning in the professions*. Jossey Bass.

SFS. (2010:800). *Skollagen*. Ministry of Education and Research.

Shulman, L. S. (1987). Knowledge and teaching: Foundations of the new reform. *Harvard Educational Review*, *57*(1), 1–21. doi:10.17763/haer.57.1.j463w79r56455411

Sjöberg, L. (2010). "Same, but different." En genealogisk studie av den 'goda' läraren, den 'goda' eleven och den 'goda' skolan i svenska lärarutbildningsreformer 1940–2008 ["Same, but different." A genealogical study of the 'good' teacher, the 'good' learner, and the 'good' school in Swedish teacher education reforms 1940–2008]. *Educare*, *1*, 73–99.

Sjöberg, L. (2018). The Swedish primary teacher education programme: At the crossroads between two education programme traditions. *Education Inquiry*, *41*(5), 604–619.

SKL. (2018a). *Fakta förskola* [Facts preschool]. Swedish Association of Local Authorities and Regions. https://skl.se/skolakulturfritid/forskolagrundochgymnasieskola/forskolafritidshem/forskola/faktaforskola.3292.html

SKL. (2018b). *Skolans rekryteringsutmaningar: Lokala strategier och exempel* [The recruitment challenges of school: Local strategies and examples]. Swedish Association of Local Authorities and Regions. https://webbutik.skl.se/sv/artiklar/skolans-rekryteringsutmaningar-.html

Skolverket. (2019). *Forskning och utvärderingar.* [Research and evaluations]. https://www.skolverket.se/skolutveckling/forskning-och-utvarderingar

Sollerman, S. (2019). *Kan man räkna med PISA och TIMSS?: Relevansen hos internationella storskaliga mätningar i matematik i en nationell kontext* [Can you count on PISA and TIMSS?: The relevance of international large scale measurement of mathematics in a national context]. Doctoral dissertation, Stockholm University, Department of Mathematics and Science Education.

Statens Offentliga Utredningar (SOU). [Swedish Government Official Reports]. (1965). *Lärarutbildningen: 1960 års lärarutbildningssakkunniga IV:I* [Teacher education: The 1960 expert rapport for teacher education IV:I]. Esselte AB.

SOU. (1978):86. *Lärare för skola i utveckling: Betänkande av 1974 års lärarutredning* [Teachers for a developing school: Report from the 1974 teacher inquiry] *(LUT 74)*. Gotab.

SOU. (1999):63. *Att lära och leda—en lärarutbildning för samverkan och utveckling* [To learn and to lead—A teacher education for collaboration and development]. Nordstedts Tryckeri AB.

SOU. (2008):109. *En hållbar lärarutbildning: Betänkande av utredningen om en ny lärarutbildning* [A sustainable teacher education: Report of the inquiry into a new teacher education] *(HUT07)*. Fritzes.

Spaiser, V., Hedström, P., Ranganathan, S., Jansson, K., Nordvik, M. K., & Sumpter, D. J. (2018). Identifying complex dynamics in social systems: A new methodological approach applied to study school segregation. *Sociological Methods & Research*, *47*(2), 103–135.

Sumpter, L. (2018). "Look at Finland." In O. Kwon, S. Lee, & M. Park, (Eds.), *Proceedings of the 2018 International Conference of the Korean Society of Mathematics Education: The professional development of all teachers of mathematics* (pp. 73–79). Seoul National University of Education.

Sundsdal, E., & Øksnes, M. (2015). Til forsvar for barns spontane lek [In defence of children's spontaneous play]. *Nordisk tidskrift for pedagogikk og kritikk* [Nordic Journal of Pedagogy and Critique], *1*, 1–11.

Svensson Källberg, P. (2018). *Immigrant students' opportunities to learn mathematics: In(ex)clusion in mathematics education*. Doctoral dissertation, Department of Mathematics and Science Education, Stockholm University.

Swedish Council for Higher Education. (1993/2018). *The Higher Education Ordinance* of 1993, including amendments up until 2018 (SFS 2010:541). Swedish Council for Higher Education. Translated to English by Swedish Council of Higher Education, 2019. https://www.uhr.se/en/start/laws-and-regulations/Laws-and-regulations/The-Higher-Education-Ordinance/

Swedish Council for Higher Education. (2019). *Bolognaprocessen—det europeiska området för högre utbildning* [The Bologna process—the European area for higher education]. https://www.uhr.se/internationella-mojligheter/Bolognaprocessen/

Swedish National Agency for Education. (2011). *Curriculum for the upper-secondary school* (revised 2018). Skolverket.

Taub, D. (2019). *Improving theory-practice connections in pre-service mathematics teacher education: A systematic review*. Masters dissertation, Department of Mathematics and Science Education, Stockholm University.

Utsedde Kommitterade [Appointed Commissioner]. (1893). *Förslag till ändrade bestämmelser rörande Undervisningsprof för lärarebefattningar vid rikets allmänna läroverk jämte därmed i samband stående frågor* [Suggested changes regarding the national teaching demonstration exams for public grammar schools with all its related issues]. Ivar Hæggströms boktryckeri.

Vulic, D. (2017). *En analys av likhetstecknets introduktion i svenska läromedel* [An analysis of the introduction of the equal sign in Swedish textbooks]. Independent thesis Advanced level (professional degree). Faculty of Science. Stockholm University.

Wågsås Afdal, H., & Nerland, M. (2014). Does teacher education matter? An analysis of relations to knowledge among Norwegian and Finnish novice teachers. *Scandinavian Journal of Educational Research*, *58*(3), 281–299. DOI: 10.1080/00313831.2012.726274

Wing, J. M. (2006). Computational thinking. *Communications of the ACM*, *49*(3), 33–35.

Wyatt-Smith, C., & Klenowski, V. (2013). Explicit, latent and meta-criteria: Types of criteria at play in professional judgement practice. *Assessment in Education: Principles, Policy & Practice*, *20*(1), 35–52.

CHAPTER 3

TEACHER TRAINING IN MATHEMATICS IN FRANCE

Nadine Grapin
André Revuz Laboratory of Didactics, University Paris-Est-Creteil

Nathalie Sayac
André Revuz Laboratory of Didactics, University Rouen Normandie

In this chapter we explain how mathematics teacher training programs are structured and developed in France. We begin by describing the organization of the French school system (from primary school to university) and the ways in which teacher training programs are structured are presented. Although mathematics is taught by several types of teachers according to the level (primary school teachers, who teach several subjects, and secondary school teachers, who teach only mathematics), the path for becoming a teacher is similar: they must first pass a recruitment examination (4 years after Grade 12—equivalent to the 1st year of a Master's degree), after which they become trainee teachers for one academic year. Extracts of tests are then used to describe the content of these exams and how the Graduate Schools of Teaching and Education are organized for the training year, which follows recruitment. Two specificities of training in France are the focus of this chapter: dual training (trainee teachers who work alternatively at school and in training) and "masterization" (most teachers have the grade of Master—five years after grade 12). The chapter concludes with a discussion about advantages and disadvantages of teacher training practices in France and how mathematics teacher training is likely to evolve in the years to come.

International Perspectives on Mathematics Teacher Education, pages 49–74.
Copyright © 2021 by Information Age Publishing
All rights of reproduction in any form reserved.

INTRODUCTION

In France, the school system is founded on several general principles: education is compulsory for all children between the ages of 3[1] and 16; is free until the end of upper secondary level; and is secular. In the first section of this chapter is a description of the country-specific context of schooling in France, specifically the organization of the French school system from primary school to university and its management by local authorities. In the second section we describe the structure of teacher training programs and the different types of teachers in France. This is followed by a description of the content of competitive teacher recruitment examinations and how teachers are prepared to teach mathematics, distinguishing between primary and secondary school teachers. The chapter concludes with a discussion of the experiences of mathematics teachers following their initial preparation.

SCHOOL EDUCATION IN FRANCE

We describe the French school system and its management. For information about the French national curriculum for mathematics, see Gueudet et al. (2017).

Primary and Secondary School

Beginning with the 2019–2020 academic year, pupils will start compulsory education at the age of three at nursery school, which they will attend for three years. Then at the age of six, they begin elementary school and progress to lower secondary at age 11, and then to upper secondary at age 15 (See Figure 3.1). Cycle 3 represents the transition between primary and secondary school. At the end of ninth grade, pupils can sit exams to obtain the DNB *(Diplôme national du brevet)*, which partially validates the common core of knowledge, skills, and culture (Décret, mars 2015) and the ability to thrive in a social and professional environment.[2] After the orientation cycle (grade 10), students choose among the vocational, technical, or general route and prepare for their specific diploma *(baccalaureat[3])* at the end of each route. The students' choices for further education

---

[1] Education from the age of 3 years old has been compulsory since the "School of Trust" law was passed in June 2019. Prior to this, education was compulsory from the age of 6. Although in practice most 3-year-olds already attended nursery school, this recent legislation gives more importance to the role of these first years of education.

[2] Students can also sit another exam, the CFG *(Certificat de Formation Générale)*. This exam especially concerns students with learning difficulties; it can be obtained through continuous assessment throughout the completion of mathematics and French courses and after having passed an oral examination. Since 2018, the DNB has also been based on continuous assessment as well as two written tests (a first for mathematics and sciences and a second for French, history-geography, civics, and ethics) and an oral test (for which the student presents a project developed over the three previous years).

[3] The French "Baccalauréat" is the diploma that students take at the end of secondary school. It is different from the university "Baccalaureate" (a term used in several Anglo-Saxon countries), which

	School Type	Age	Grade	Cycle
PRIMARY SCHOOL (ages 3–11)	Nursery School (compulsory since September 2019)	3–4	First year	1st cycle (early learning)
		4–5	Second year	
		5–6	Third year	
	Elementary School (compulsory)	6–7	First grade (cours préparatoire—CP)	2nd cycle (basic learning)
		7–8	Second grade (cours élémentaire 1ère année)	
		8–9	Third grade (cours élémentaire 2nde année)	
		9–10	Fourth grade (cours moyen 1ère année)	3rd cycle (consolidation)
		10–11	Fifth grade (cours moyen 2ème année)	
SECONDARY SCHOOL (ages 11–18)	Lower Secondary School—College (compulsory)	11–12	Sixth grade (sixième)	
		12–13	Seventh grade (cinquième)	4th cycle (further learning)
		13–14	Eighth grade (quatrième)	
		14–15	Ninth grade (troisième)	
	Upper Secondary School—Lycée (compulsory up to age 16)	15–16	Tenth grade (seconde)	Orientation cycle
		16–17	Eleventh grade (première)	Baccalaureate route
		17–18	Twelfth grade (Terminale)	

FIGURE 3.1.   Organization of the French School System by Age

depend on their interest in a specific field or career as well as on the decision of the class council[4] regarding their knowledge and abilities.

At the beginning of the 2017 school year, approximately 12 million students attended French schools: 6,783,300 in primary schools (Ministère de l'Éducation Nationale [MEN], 2018, p. 65) and 5,629,800 in secondary schools (MEN, 2018, p. 84). Around 15% of primary school students and 22% of secondary school students are in private establishments, for religious or pedagogical reasons.

For each level, the curriculum is national and described by cycle. The Common Core of Knowledge, Skills, and Culture identifies, in combination with the curriculum, the knowledge, skills, values, and attitudes necessary for a student to be successful in their schooling, personal life, and life as a future active citizen. Students are expected to know and master the Common Core by the end of their compulsory schooling, between the ages of 6–16 years old (Décret, mars 2015).

In nursery schools, emphasis is on the acquisition of language and its development, the discovery of words and numbers, and learning to live together and with others. In elementary school, students are taught several subjects and disciplines. They learn French, a foreign language (English in most cases), art, music, physical education and sports, moral and civic education, history-geography, and mathematics. Since 2008, all students are taught 24 hours per week (864 hours per year), divided into eight or nine morning or afternoon sessions. Over these hours, 180 hours are dedicated to mathematics (5 hours per week), including all topics (e.g., number sense, geometry, solving problems); every day, there is a moment of mental calculation.

In lower secondary school, students attend 26 compulsory teaching hours per week, allocated to disciplinary teaching, especially mathematics, with 4.5 hours per week in grade 6 and 3.5 hours in grades 7–9. In addition to disciplinary teaching, during cycle 4 (grades 7–9), students complete interdisciplinary practical lessons. The purpose of these is for students to contextualize and use their learning to complete collective projects chosen by teachers in each lower secondary school. Examples include designing and developing a wind turbine model, discovering a certain profession, or designing a magazine about steam engines. We can assume there are fewer hours of mathematics in grades 7–9 compared with grades 1–5 because students will apply mathematics in the context of their projects. Figure 3.2 contains a summary of the distribution of compulsory mathematics lessons in comparison to other lessons between grades 1 and 9.

Until 2018–2019, in the first year of upper secondary school (grade 10), all students were taught general classes such as French, history and geography, mathematics, etc. for 28.5 hours a week. This includes 4 hours of mathematics lessons.

---

students obtain at the end of a 4-year university degree.

[4] The class council meets three times per year with all the teachers of the class to give an opinion on the orientation and the work of the students.

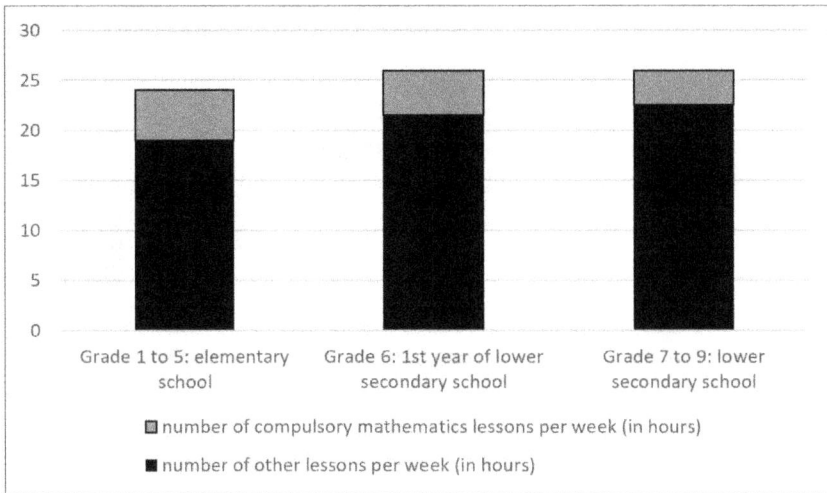

FIGURE 3.2.   Distribution of General and Mathematics Lessons Per Week

For their last two years at upper secondary school, students choose between general or technological routes. With the general baccalaureate route, students take classes centered around one of three specialized programs: Literature (L); Economics and Social Sciences (ES); or Mathematics and Experimental Science (S). The number of compulsory mathematics lessons per week varies greatly across these specialized programs: 3 hours in grade 11 and 4 hours in grade 12 for the ES program; 4 hours in grade 11 and 6 hours in grade 12 for the S program; and no compulsory mathematics courses for the L program, although students in that program can choose optional mathematics courses if they wish to do so (3 hours in Grade 11 and 4 hours in Grade 12). Students can also choose optional mathematics lessons in the ES and S programs in Grade 12 (1.5 hours in the ES program and 2 hours in the S program). Figure 3.3 contains a summary of this distribution.

The goal of the general routes (L, ES, S programs) is to prepare students for higher education after secondary school. In contrast, the technological route is to prepare them for advanced technological studies in the fields of science and technology. There are several technological routes from which to choose (e.g., Laboratory Science and Technology; Music and Dance), but all include compulsory mathematics courses at grades 11 and 12; the number of hours varies between 2 and 4 hours per week in grades 11 and 12, depending on the program.

A new baccalaureate is planned to be rolled out in 2021. Technological routes will not change, but general routes will disappear and there will be no more specialized programs. In these cases, students will have to choose *specialties* begin-

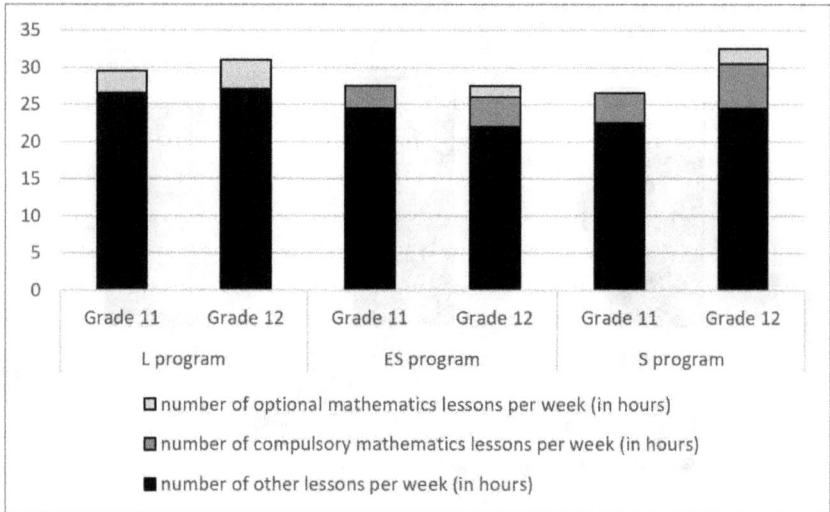

Chart shows bars for Grade 11 and Grade 12 across L program, ES program, and S program, with a vertical axis from 0 to 35 in increments of 5.

Legend:
□ number of optional mathematics lessons per week (in hours)
▨ number of compulsory mathematics lessons per week (in hours)
■ number of other lessons per week (in hours)

FIGURE 3.3. Distribution of Mathematics Lessons Per Week for the Different Programs

ning at grade 11. These specialties can be arts, mathematics, foreign languages and modern literature, physics and chemistry, among others. Depending on the specialty, students could pass their exit exams without learning any mathematics during the last two years of secondary high school. In this new program, students take only four exams, including one oral, whereas they took around ten exams in the previous program. One of these four exams may be mathematics, if students select mathematics as specialty.

University Structure

The structure of higher education in France is complex and largely goes beyond the scope of the present chapter.[5] Therefore, Figure 3.4 contains merely a summary of the structure of teaching at a university within the LMD (*Licence—Master—Doctorate*) system. This system standardizes the levels and organization of university degrees in most countries in the European Union.

It is necessary to have successfully passed a high school *baccalaureat* (either general, technological, or professional) or an equivalent qualification to enroll at a French university. Teachers and researchers provide three types of classes: lectures, tutorials, and practical work (or lab work for certain fields).

[5] For further information about the French higher education system, visit http://www.enseignement-sup-recherche.gouv.fr/.

Doctorate (PhD Thesis)		
8		
7		3rd University Cycle
6		
Diploma: Master		
5	2nd year of master (M2)	2nd University Cycle
4	1st year of master (M1)	
Diploma : Licence		
3	3rd year of license (L3)	
2	2nd year of license (L2)	1st University Cycle
1	1st year of license (L1)	
Baccalaureat (Grade 12)		

FIGURE 3.4.   Structure of the University System

School System Management

Curricula and payment of public teachers and educational staff are centrally organized by the Ministry of National Education, Youth and Sports. Educational policy for primary and secondary school is formulated for the whole country. For this policy to be applied at a local level, a Rector, who is a representative of the Ministry of Education, is designated in each academic district. There are 26 metropolitan academic districts and 4 overseas academic districts.

GENERALITIES OF TEACHER TRAINING IN FRANCE

The competitive teacher recruitment examination is the core of teacher training in France. For the purpose of this chapter, three periods of teacher training in France are discussed in relation to the exam:

- From high school graduation to the competitive teacher recruitment examination;
- After passing the teacher recruitment examination; and
- When trainee-teachers become fully-fledged teachers.

It should be noted that the professional skills described in the reference document (Arrêté, juin 2013) are the main subject of the training provided during these three periods. To gain a better understanding of the specificities of teacher training programs for mathematics, we first describe the different types of mathematics teachers and the reference document of teachers' knowledge and skills. We then present the first period of the training course, differentiating between future primary school teachers and secondary school teachers.

DIFFERENT TYPES OF FRENCH TEACHERS

In 2018, there were approximately 334,000 primary school teachers and 402,000 secondary school teachers, of which 46,500 were mathematics teachers. In elementary schools, teachers (primary school teachers: *professeurs[6] des écoles)* teach all subjects to their classes (mathematics, as well as sciences, French, arts, and others). In contrast, lower and upper secondary school teachers (in general and technological routes) are principally specialists in one subject. As such, mathematics is taught by mathematics teachers *(professeurs de mathématiques)*. In the vocational route, teachers who teach mathematics also teach physics. These clarifications are important for understanding the different ways in which teacher education is organized according to the different levels of teaching (primary or secondary school).

In primary school, teachers teach 24 hours a week. In secondary school, there are two classifications of teachers based on the recruitment examination they passed: specialist teachers *(professeurs agrégés)*; and certified teachers *(professeurs certifiés)*. The exam for becoming a specialist teacher requires the candidate to demonstrate a higher level of mathematical skills than that for becoming a certified teacher. Generally (but not necessarily), specialist teachers teach in upper secondary school or at a higher level, in classes where the program for the discipline is superior (for example, in the S program). Mathematics teachers teach 15 hours (specialist professors) or 18 hours (certified professors) per week. It should be noted that specialist and certified professors teach the same program, but specialist professors are better paid than certified professors for less teaching time. These differences demonstrate the elitist nature of the French school system.

Reference Document of Teachers' Skills and Competencies

In France, a reference document, the Reference Framework for Professional Skills for Professions in Teaching and Education (Arrêté, juin 2013), sets out the skills that all teachers (whatever the subject or level taught) need to develop and defines common aims, knowledge, and culture. This framework is the same for all teachers, at public or private schools, primary or secondary.

Teachers begin to acquire these skills during their initial training and continue to develop them throughout their careers (Arrêté, juin 2013). In this document, teacher skills are divided into several categories about education, instruction, and teaching, directly linked to disciplinary and didactic knowledge (for the present purpose, mathematical knowledge).

---

[6] In North America, the word "professor" tends to refer to university-level teachers. In France, however, the word professeur is used for a teacher (from nursery to grade 12 or even in higher education). The term Professeur des Universités is also used for a certain category of teacher-researcher.

First, skills serve to remind teachers that they are civil servants who contribute to public service education. In this context, they must share the values of the Republic and operate in accordance with the fundamental principles of the school system and within the regulatory school framework. Teachers are also educators to support students' success. They must have knowledge of child psychology and learning processes and must also consider the diversity of their students (for example, students with disabilities, those at risk of dropping out, or those with special educational needs). Each teacher must master the French language to communicate and be able to use a foreign language in professional situations. The ability to integrate digital literacy is necessary as teachers have to use technologies to communicate and contribute to their students' education in terms of using technologies in a responsible way. They are expected to integrate technologies into their teaching, such as geometry software and spreadsheets. They must also use technologies outside of the classroom, such as using online resources on digital workplace platforms for designing lessons or sharing documents with parents, students, or administrators. Teachers are defined as actors of the educational community: they have to collaborate in a team, as well as engage with their students' parents. They must also engage in a personal and collective process of professional development.

Common skills are defined for all teachers: mastering disciplinary knowledge and how to teach the subject, and mastering French in a teaching context. Relying on professional skills, teachers must also design, implement, and host teaching and learning situations, taking into account the diversity of their students. They must establish and oversee a modus operandi within the class, supporting learning and socialization, and assessing students' progress and achievements.

The "Masterization" of Teacher Training

Before outlining the teacher training program for elementary and then middle and high school teaching levels, we conclude this section with a historical overview of teacher training in France, to explain how universatization and professionalization have changed the system for recruiting and training teachers (Cornu, 2015). This historical perspective focuses on several main ideas, including the role of universities in teacher training, the importance of both academic and practical teacher training, the development of teaching skills, and the role of research in teacher training.

Since 2013, initial teacher training has been provided by specific university departments (*ESPE:*[7] *Ecole Supérieure du Professorat et de l'Education*—Graduate Schools of Teaching and Education) affiliated with one or several universities. An ESPE is a specific training center developed to help individuals become successful teachers. There are 32 ESPEs located throughout the French academic districts. Staff at ESPEs also take part in disciplinary and educational research and organize training activities for teaching staff in primary and secondary education

---

[7] Since September 2019, the ESPE has been renamed INSPE (Institut National Supérieur du Professorat et de l'Education, or National Higher Institute of Teaching and Education).

as well as for other education personnel. Prospective teachers could prepare to take one of the recruitment examinations in an ESPE, which is highly encouraged, but not compulsory. They also could prepare by themselves or with a private institute. Those who prepare for one of the different competitive exams in an ESPE are involved in a training program as part of a MEEF[8] Master's degree. They have early contact with students, spend time in classes (observations during the first year of the MEEF Master's degree, sole responsibility during the second year), and receive support, including tutoring by other qualified teachers. The training program for teachers and educational professionals includes a core curriculum offered to all trainee teachers (primary and secondary school teachers alike).

From High School Graduation to the Competitive Teacher Recruitment Examination

Teacher recruitment exams are held at the end of the first year of a Master's degree (M1—4 years after Grade 12). Students can enter the teacher recruitment exam if they hold any Master's degree, but parents of 3 children or top-level athletes can be dispensed from having a Master's degree.[9] Thus, most candidates have a *Licence* (3 years after Grade 12), usually in mathematics for secondary school teachers but in any discipline (e.g., psychology, education sciences, arts) for primary teachers. Most future middle and high school mathematics teachers thus deepen their knowledge of mathematics during this period. However, future elementary school teachers have not necessarily followed a course with mathematical or even scientific content.

After obtaining a *Licence*, each student can enroll in an ESPE in a Master's degree in the fields of education and training (MEEF), but this is not compulsory. During this first year, students can complete an observation internship in an elementary or middle or high school before passing a competitive teacher recruitment exam. Most Master's degree courses focus on specific subjects, and future mathematics teachers focus on mathematical content. However, future elementary school teachers take courses in a variety of disciplines (French, arts, sports, sciences, etc.) including mathematics, with a variable number of hours.

COMPETITIVE RECRUITMENT EXAMINATIONS

In this section we describe the content of the different recruitment examinations. First, we discuss the examinations for future primary school teachers, followed by a discussion of examinations for secondary school teachers (certified and specialist teachers). Primary school and secondary school teachers in the French public-school system are State Civil Servants and are recruited after completing a two-year

---

[8] MEEF : Métiers de l'Enseignement, de l'Education et de la Formation (Teaching, Education and Training Profession)

[9] This is a specific condition granted in France to all competitive examinations for the recruitment of state officials.

Master's degree program (5 years after grade 12) and passing a competitive entry exam based on the grade level they wish to teach (primary or secondary school).

Primary School Exam

To teach in primary schools, candidates take the competitive primary teacher recruitment examination (CRPE[10]) organized by the regional educational authority. This recruitment competition is divided into two parts: an eligibility test and an admission test. The aim of all the tests is to assess candidates' abilities with regard to the disciplinary, scientific, and professional dimensions of teaching and consists of two written eligibility tests (French and mathematics) and two oral admission tests (vocational training and interview from a file developed during their preparation).

The duration of the written tests in French and in mathematics is 4 hours for each test. The aim of the test in mathematics, which consists of three parts, is to assess the candidate's mastery of the disciplinary and didactic knowledge required to teach mathematics in primary school. The test is scored out of 40 points: 13 for the first part, 13 for the second, and 14 for the third.

The first part of the test consists of a problem relating to one or more areas of the elementary school or lower secondary school programs, or to elements of the common base of knowledge, skills and culture, to assess the candidate's ability to search, extract and organize useful information. Figure 3.5 provides an example of such a task.[11]

The second part of the test consists of independent mathematics exercises, complementary to the first part, to verify the candidate's knowledge and skills in different areas of the elementary school's or lower secondary school's programs. These exercises may be proposed in the form of multiple-choice questions, built-in answer questions, or error analyses of potential student responses to determine their origins. Figure 3.6 provides an example of such a task.[12]

The third part of the test consists of an analysis of documents contained in a folder to assess the candidate's ability to master the concepts presented in teaching situations in mathematics. The documents could be one or more pieces of mathematics teaching materials, selected from primary school curricula for pupils or teachers (textbooks, teaching materials), and responses from pupils.[13] Figure 3.7 provides a sample exercise.

---

[10] *Concours de Recrutement des Professeurs des Ecoles:* Recruitment Examination for Primary School Teachers

[11] For more information about this assessment see http://www.devenirenseignant.gouv.fr/cid98675/sujets-des-epreuves-ecrites-conseils-des-jurys-des-concours-recrutement-professeurs-des-ecoles.html which provides information about this assessment in untranslated French.

[12] For more information about this assessment see http://www.devenirenseignant.gouv.fr/cid98675/sujets-des-epreuves-ecrites-conseils-des-jurys-des-concours-recrutement-professeurs-des-ecoles.html which provides information about this assessment in untranslated French.

[13] For more information about this assessment see http://www.devenirenseignant.gouv.fr/cid98675/sujets-des-epreuves-ecrites-conseils-des-jurys-des-concours-recrutement-professeurs-des-ecoles.html which provides information about this assessment in untranslated French.

Exercise: 3 Squares

The figure below represents 3 squares whose side measurements, in centimeters, are respectively 3 cm, 4 cm, and 5 cm. The two small squares are grey, the third is white.

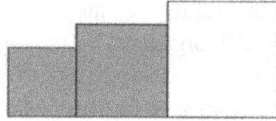

1. Check that the sum of the areas of the two grey squares equals the area of the white square.
2. Claude says: "if you have 3 squares obtained from the previous question (as in Figure 1 below), then the *ABC* triangle is a right triangle."
   Is Claude's statement true or false? Justify your answer.

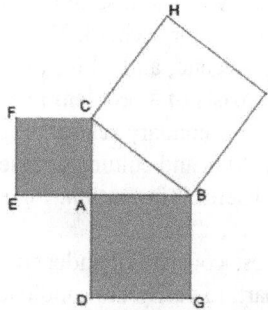

Figure 1

3. With the same squares, Dominique states: "In Figure 2 below, the exact lengths, in centimeters, of the *MN* and *IJ* segments are decimal numbers."

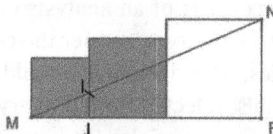

Figure 2

Is Dominique's statement true or false? Justify your answer.

FIGURE 3.5. Example Task From the First of the Four Proposed in the First Part of the Primary Test 2019 (group 1)

Exercise 3

You have several solid cubes (not hollow) with a 5 cm edge. Some are made of iron, others are made of nickel. The density of iron is 7860 kg/m³, that of nickel is 8900 kg/m³. If you choose a cube, weigh it and find that its mass is 1110g, is this cube made of iron or nickel?

FIGURE 3.6. One of the Four Exercises Proposed in the Second Part of the 2019 Primary Test (group 2)

The success of the first part of the recruitment examination (as well as the following) depends on candidates' answers, as this is a competitive exam. For each academic district, a number of teaching posts are offered based on the academic needs. This means that it is easier to pass the examination in certain academic districts than in others. It is arguably unfortunate that it is in the academic districts where teaching is more difficult (because the social and economic context is more difficult) that the recruitment examination is easier to pass. This means that, in

---

Exercise 3

During a calculation lesson, a grade 5 teacher proposes the following calculation:
$12.47 + 2.7$

Here are the workings of four students:

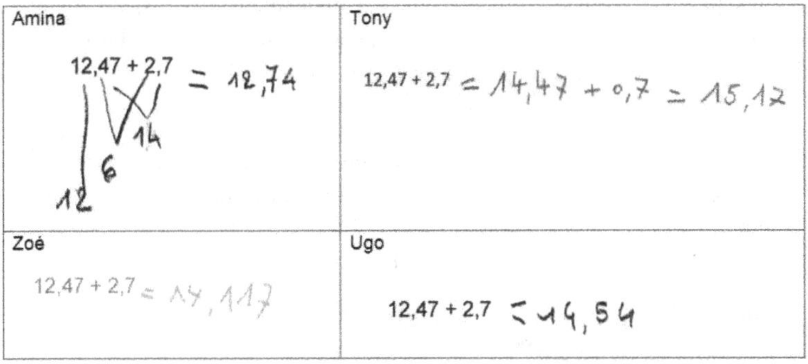

1. How is a single line calculation a complementary method to mental calculation?

2. For each student, explain the procedures used and analyze any errors they made.

3. What support could the teacher suggest to Zoe to help to correct her error?

FIGURE 3.7. The Third of Three Tasks Proposed in the Third Part of the 2019 Primary Test (group 2)

these academic districts, candidates could succeed with a very low score, especially in mathematics. When a candidate has passed the exam in one academic district, he is compelled to teach in this district and can only move to another district after a few years.

Secondary School Exam

Secondary school teachers are specialized in one discipline. They are recruited in competitive entry exams at the Master's degree level (5 years after grade 12) and become public servants. Certified teachers hold the Certificate of Proficiency in Secondary Education (*CAPES: Certificate of Aptitude for Secondary School Teaching*) in their discipline, while the specialist teachers obtain an *Agrégation* (Aggregation in English). Historically, this Aggregation was established in 1776 and the CAPES in 1950. The first is more selective and requires a higher level of mathematical knowledge. These two competitive entry exams (CAPES and Aggregation) are nationwide; the first part of the test (eligibility tests) is written and the second is oral.

Certificate of Proficiency in Secondary Education (CAPES)

The first part of the CAPES assesses the candidate's mathematical knowledge, reasoning and argumentation abilities, as well as their command of the French language. Even if candidates have a 4 year-university degree when they take this exam, the level of mathematics involved in problem solving is *licence* level (3-year university degree level). The competitive program includes, for example, linear algebra and matrix calculations:

> Linear Algebra: linear systems, the Gauss-Jordan elimination algorithm; finite dimensional vector spaces, free families, generating families, bases; linear applications; homotheties, projections and symmetries; rank of a linear application; endomorphism matrix representations. Reduction of endomorphisms and square matrix [...]

> Matrix and Determinants: matrix calculation, invertible matrix, transposition; matrix and linear applications, base changes; equivalence, similarity; determinants of a square matrix, of an endomorphism and of a finite dimensional vector space. (Ministère de l'Education Nationale, 2019a)

For example, the first part of the written exam of the 2019 session began with the questions shown in Figure 3.8.[14]

In the second part of the CAPES eligibility test, candidates must solve several secondary school-level mathematical problems. These questions aim to assess whether future teachers have solid mathematical knowledge about the content of secondary school teaching. Beyond the scientific qualities of the candidate,

---

[14] For more information about this assessment see https://media.devenirenseignant.gouv.fr/file/Capes_externe/13/7/p2022_capes_ext_mathematiques_1404137.pdf which provides information about this assessment in untranslated French.

I. Question in a mathematics lesson. Given a real number θ non congruent with 0 modulo 2π, Ω a point of $\mathcal{P}$ and $\vec{u}$, a vector of $\mathcal{P}$. The affix of Ω is noted ω and the affix of $\vec{u}$ is noted $z_{\vec{u}}$.

Given a point M of $\mathcal{P}$ whose affix is z.

    a.   Find the affix z' of the image of M under the translation $t_{\vec{u}}$ (translation by $\vec{u}$).

    b.   Find the affix z'' of the image of M under the rotation $r_{\Omega,\theta}$ (rotation of center Ω and of angle θ).

II. Given a a complex number with modulus 1 and b a complex number. Consider function f from $\mathcal{P}$ to $\mathcal{P}$, that maps each point whose affix is z to the point whose affix is $az + b$.

1. Prove that, if $a = 1$, then f is a translation, the translation vector will be specified.

2. In this question, we consider $a \neq 1$.

    a.   Prove that f has a single fixed point Ω whose affix ω will be specified.

    b.   Prove that the image of the point M under f is the point whose affix is $a (z - \omega) + \omega$.

FIGURE 3.8.   Task From the Written Exam for CAPES (2019)

this test also assesses their aptitude to be placed in a professional position. As an example, Figure 3.9 contains the first questions of the 2019 session.[15] Natural logarithm is taught in Grade 12 in the ES and S programs, but logarithm to base a is not.

Oral admission tests include a conversation with a jury and are organized in two parts. First, the candidate presents a detailed plan of a study theme (the mathematics involved is upper secondary school level) and explains parts of this plan (according to the jury's questions). The list of study themes is published on an official website ((Ministère de l'Education Nationale, 2019b). For the 2020 session, there were 39 themes, for example, "discrete random variables," "metric and angular relations in a triangle," or "Primitives—Integrals."

The second part of the admission tests is based on several documents provided by the jury (e.g., extracts from the official curriculum or textbooks, student work) dealing with a theme taken from the mathematics programs for secondary school. Candidates' answers to the questions are used to assess their teaching skills and abilities. During this conversation, the jury can also assess how candidates share the values of the Republic of France (freedom, equality, fraternity, secularism, and absence of all forms of discrimination) and how they take their students' difficulties into account.

---

[15] For more information about this assessment see https://capes-math.org/ which provides information about this assessment in untranslated French.

Part A – base *a* logarithm

Reminder: we call *logarithm* all functions f defined on $]0 ; + \infty[$, differentiable, such that:

- there is a real number a ($a \neq 0$) such that for any positive real number x, $f'(x) = \frac{a}{x}$.

- $f(1) = 0$

I.   Given a a real number, $a \neq 0$. Prove that there is a single logarithm, wherein we can denote f_a, such that, for any real number $x > 0$, $f_a'(x) = \frac{a}{x}$. If $a = 1$, we use notation ln (Napierian logarithm).

II.   For any real number a ($a \neq 0$), give the expression of f_a using ln.

III.   Prove that, for any real number a ($a \neq 0$), and for any real numbers $x, y > 0$:
$f_a(xy) = f_a(x) + f_a(y)$.

    Indication: we could analyze the function defined by $x \rightarrow f_a(xy)$.

IV.   Prove that for any real number $x > 0$: $f_a\left(\frac{1}{x}\right) = -f_a(x)$.

FIGURE 3.9 .   Secondary-Level Task From the CAPES Exam (2019)

Agrégation

The *Agrégation* is designed in the same way as the CAPES but is more focused on mathematical skills. The content of the two written admissibility tests involves specific mathematics knowledge at a higher level than CAPES (Master's level for the Aggregation and Licence level for the CAPES). Candidates must solve mathematical problems about general mathematics (for the first test) and analysis and probability (for the second). As an example, Figure 3.10 contains the first part of a mathematical problem about general mathematics for the 2019 exam (Ministère de l'Education Nationale, 2019c). From this example alone, we can see the difference between the CAPES and the Aggregation in terms of the type of mathematics involved in the problems. Algebraic elements, rings, etc. are not expected for the preparation of CAPES. N.B. all notation used in the text below are introduced before the problem.[16]

There are three oral admission tests. The first is about *algebra* and *geometry*, the second about *analysis* and *probability*, and the last is about *modeling*. For the two first oral exams, candidates must present a detailed plan of a study theme in the same way as for the CAPES, but the mathematics involved for this is at the *Licence* level. For example, themes could be "space vector modulations," "bilinear or quadratic forms in a vector space," or "Fourier transformation."

The final *Modeling* test is specific. Candidates must choose between two texts (around 5–6 pages) based on concrete problems, such as, "We are interested in the

---

[16] For more information about this assessment see https://agreg.org/ which provides information about this assessment in untranslated French.

I. **Preliminary exercises**

1. Let $B \in \mathbf{Z}[X]$ a unitary polynomial and $A \in \mathbf{Z}[X]$. Prove that there is $Q, R \in \mathbf{Z}[X]$ such that $A = BQ + R$ with deg R < deg B or $R = 0$.

    Indication: we can demonstrate this through its recurrence on the degree of A.

2. **The $\mathbf{Z}[j]$ ring.** We note $j = e^{\frac{2i\pi}{3}}$.

    (a) Prove that j is an algebraic element on $\mathbf{Q}$ and specify its minimal polynomial.

    (b) Prove that $\mathbf{Z}[j] = \{a + bj, (a, b) \in \mathbf{Z}^2\}$.

    For any complex number z, let $N(z) = z\bar{z} = |z|^2$.

    (c) Prove that, for any $z \in \mathbf{Z}[j]$: $N(z) \in \mathbf{N}$. **Calculate** that, if $z \in \mathbf{Z}[j]$ is invertible, then $N(z) = 1$ and $\mathbf{Z}[j]^\times$ has 6 elements which will be specified.

    (d) Given $x \in \mathbf{Z}[j]$ and $y \in \mathbf{Z}[j] \setminus \{0\}$. Determine one element $q \in \mathbf{Z}[j]$ such that $N\left(\frac{x}{y} - q\right) < 1$. Conclude that the $\mathbf{Z}[j]$ ring is Euclidian.

FIGURE 3.10.   Sample Problem From the 2019 Aggregation Exam

use of numerical matrix analysis methods in the context of bibliographic database management. Key words: linear algebra. Matrix elements. Least squares." This text is followed by the description of a mathematical model to use in order to solve the problem, with theoretical aspects and suggestions to help. Candidates must prove some theorems involved in the model, explain how to use this model to solve the problem, and are encouraged to use mathematical software to illustrate their explanation. They have 40 minutes in which to give their presentation and 25 minutes to answer the questions of the jury. This oral presentation, therefore, requires the ability to synthesize and show initiative, but also gives the candidate the opportunity to gain some perspective about their own mathematics knowledge.

Mathematical skills can be easily assessed in competitive entry exams. However, few professional skills can be assessed in this way, such as the candidate's analysis of their students' needs.

INITIAL AND ONGOING TEACHER TRAINING

In this part, we describe initial and ongoing training from which teachers can benefit throughout their career. We explain how this is done at two times: the year after passing a teacher recruitment exam (initial training); and thereafter, when students become fully-fledged teachers (ongoing training).

After Passing a Teacher Recruitment Exam

After passing the teacher recruitment exam, successful students become trainee teachers. During a school year corresponding to their second year of a MEEF

Master's degree, they split time between a school where they teach and an ESPE where they are trained. They are paid by the Ministry of Education as probationary officials during this period of dual training (*formation en alternance*). Colmant (2015) explains that ESPE programs comprise "approximately 1000 hours of vocational training, divided into three parts: learning (45%), practice in schools (40%) and personal work (15%). Educating future teachers in school disciplines totaled 450 hours."

Primary School Teachers

For half of the year (every second week or 2 ½ days per week), trainee primary school teachers are responsible for a classroom.[17] Each trainee teacher is accompanied in the field by two tutors: one is generally a full-fledged, qualified teacher who works in the same school district as the trainee, and the second is an ESPE educator. These two tutors could exchange information about the trainee, but this is rare. Nevertheless, both produce a report and assess the trainees separately at the end of the year. When they are at school, teacher trainees are considered as regular teachers: they are responsible for teaching in one school class (from nursery school to grade 5). They teach all subjects, including mathematics.

The training courses at ESPE are varied. Some are dedicated to developing professional skills, others are focused on subject-specific teaching methods (didactics in all disciplines taught at school), and acquisition of skills that cut across the various subject areas related to the profession of a schoolteacher. There are also courses in sociology of education, child psychology, inclusive schooling for children with special educational needs, and on emerging themes structured around the values of the Republic (e.g., teaching of secularism, the fight against discrimination).

The distribution of the training courses depends on the training program adopted by each ESPE, even if there is a national training framework defined by the Ministry of Education. For example, the distribution of courses in the Creteil ESPE for the 2nd year of the MEEF Master's degree specific to the primary school includes mastering the French language, disciplinary knowledge and teaching skills as part of professional practice (196 hours per year including (only) 36 hours of mathematics lessons[18]), knowing and understanding the daily working context of the teacher (46 hours per year), using a living foreign language in teaching situations (15 hours), and initiation to the research preparation and completion of a Master's dissertation (45 hours).

For career changers or trainees who already have a Master's degree (whatever the topic), there is a specific program with the same types of courses. But this

---

[17] When the trainee teacher is not in the class, a fully-fledged teacher or another trainee teacher teaches the class.

[18] There is no training program in mathematics, but trainers usually propose lessons on first numeric learnings, problem solving, geometry, measurement, arithmetic (mental and calculation), operating techniques, etc.

program is completed in less time and without submission of an academic dissertation, which is replaced by a scientific essay on a professional subject (which could be mathematics-based).

Secondary School Teachers

After they have passed a CAPES or Aggregation exam, trainee secondary school teachers must teach between 8 and 10 hours per week (for certified teachers) and between 7 and 9 hours per week (for specialist teachers) (by decision of 22nd August 2014) in a lower or upper secondary school with specialist teachers priority-appointed in upper secondary schools. Each trainee teacher is accompanied by two tutors: the first is generally a full-fledged qualified mathematics teacher who works in the same school as the trainee, and the second is an ESPE educator. These two tutors regularly exchange information about the trainee, but they assess them separately at the end of the year.

When they are at school, mathematics trainee teachers are considered regular teachers, even though they have a lower teaching load than others. Generally, they must teach mathematics for two or three classes, as well as manage their classes and design their lessons. The two tutors help trainees, especially at the beginning of the year, giving them advice and supporting them at the start of their professional careers.

The training courses at ESPE are varied. Some cover mathematics content (sometimes with an historical and epistemological point of view), others are about the teaching context and classroom management (how to manage difficult students, how to communicate with parents). A lot of courses concern teaching mathematics, in which trainees learn generalities about teaching skills in relation to their practice, questions, and difficulties. For example, when teacher trainees are working on learning and teaching of a specific mathematical concept at different levels, they can study the corresponding curricula, textbook excerpts, student work, classroom videos, etc. In these courses, trainees learn how to design progressions, lessons, and assessments, and a large part of such courses is devoted to the use of technology for teaching and learning. No national program exists for this training. For example, the courses in Creteil ESPE for the 2nd year of the MEEF Master's program specific to the lower and higher secondary school are as follows: mastering mathematical content and history of mathematics (42 hours per year), studying practices for the class with didactic and digital tools (99 hours per year), knowing the daily working context of the teacher (42 hours), analyzing their practice, and completion of a Master's dissertation (48 hours). Practice analysis sessions consist, for trainees, in presenting activities that they have implemented or would like to implement in class to exchange with their colleagues and trainers. The analysis of the proposed mathematical tasks and the planned scenario of the session make it possible to consider different alternatives and to understand what effectively happened (*a posteriori* analysis) compared to what was planned (*a priori* analysis). Teachers who need support, for managing their classes for example, can have specific support.

Career changers or trainees who already have a Master's degree (whatever the topic) can follow a specific program with the same types of courses, but which is completed in less time and does not require the submission of a Master's dissertation. Instead, they prepare a reflective scientific submission *(Travail Scientifique de Nature Réflexive)*. This can be written or oral and undertaken individually or in groups. It is based on a chosen mathematics class session implemented in the trainees' classes. Teacher trainees must analyze their own teaching practices and draw on bibliographical references (from mathematical didactics or sciences of education). The potential themes for this kind of work are varied, such as the use of videos in mathematics lessons, teaching the history of mathematics, or integrating students with learning difficulties.

During their training year, all trainee teachers must submit a portfolio comprised of several types of documents: analyses of specific lessons (when they have a visit from a tutor or when they use technology); analyses of assessments; or reports on specific activities (such as a school trip or school board meeting). This portfolio is one element assessed for their tenure; the other is the way they manage their classes during their internship.

Master's Dissertation

To complete their Master's dissertation, teacher trainees can select a topic of their choice linked to their practices. For example, topics for Master's dissertations can be teaching mathematics through play, using GeoGebra to help students with 3D geometry, designing a differentiated lesson in mathematics, dealing with stress in mathematics, learning mathematics with problem solving in school, etc.

During their training year, trainee teachers must submit work related to the courses to validate their MEEF or DU[19] Master's degree. Independently of the Master's degree, they must also obtain a professional tenure from the Ministry of Education. This means that, during the second year of a MEEF Master's degree, students who are also trainee teachers must pass both their Master's degree and their internship in school. This situation makes the year following their successful teacher recruitment examination inherently complex and cumbersome.

An Example of a Primary Teacher's Master's Degree Dissertation. During the 2018–2019 academic year, a trainee schoolteacher studied performance and strategies utilized by pupils from grades 2 and 3 to solve mathematical problems (Gelas, 2019). She focused on gender differences and evolution of strategies throughout both years. To do this, she designed two tests comprising 5 problems corresponding to the Vergnaud typology (1990), adapted to the two levels of class by changing numbers. She administered these tests among 59 pupils from grade 2 and 78 from grade 3 (70 girls and 67 boys). She analyzed the productions of these

---

[19] DU = University Degree

137 pupils (mathematical errors, procedures, performances) according to two pa-rameters (gender and grade). She found that grade 2 pupils use more personal procedures (iterated additions rather than multiplications, for example) than those in grade 3 but do not succeed less. She also noted that girls use more of the proce-dures expected by the teacher, but do not succeed less than boys as she suspected.

An Example of a Secondary Teacher's Master's Degree Dissertation. Dur-ing the 2018–2019 academic year, two high school trainee teachers (Alix Duval and Antonin Gelamur) analyzed their practices for their dissertation, focusing es-pecially on how, as teachers, they used games in their lessons for teaching the first years of lower secondary school. Their 1st-year secondary school students regularly played a game called "Maths Up," a take on the well-known "Time's Up" game where players have to guess famous names. In "Maths Up," students have to guess mathematical words from definitions, pictures, or gestures. Duval and Gelamur (2019) analyzed their students' learning, but also, as teachers, their teaching practices in game situations.

To explain how they designed and developed the game, they listed several "di-dactic and play variables" (Pelay, 2011, referring to Brousseau, 1998). Didactic variables were chosen to either limit or promote some specific representations of the mathematical concept to be guessed (for example, the number of words al-lowed for each guess, with or without a drawing). Play variables (like the time for guessing, the number of guesses allowed) were also defined to ensure the game retained both fun and didactic aspects.

The teachers observed that, during the games, students corrected each other and ensured that the rules were enforced. The formalization of definitions im-proved significantly and students (even those with learning difficulties) were able to easily mobilize several representations of the same concept. Concerning their own teaching practices, by becoming less and less involved as the game pro-gressed, they observed that students themselves took over to ensure the game and learning progressed. This caused the "didactic and playful contract" (Pel-ley, 2011, referring to Brousseau, 1998)—the set of rules and behaviors, implicit and explicit, between teachers and students, which linked play and learning—to change significantly. Teachers shared more responsibility with students, both in the process of validating answers and in enforcing the rules of the game.

After Students Become Fully-Fledged Teachers

Field educators[20] are in charge of the inservice training of fully-fledged teachers, which is defined by local and national education policies. The implementation of these policies is not managed by an ESPE, but by specific academic or district bodies.

---

[20] Field educators are teachers who are recognized by the institution for their high professional skills. They will usually have passed a professional examination: the CAFIPEMF (*Certificat d'aptitude aux fonctions d'instituteur ou de professeur des écoles maître formateur*) for primary school

In primary schools, teachers have 18 hours of compulsory training per year proposed by the educational board of their school district. For a few years (after the results of TIMSS 2015)[21], a national priority was given to mathematics with specific national training programs (Mathematics Strategy in 2014 and Villani-Torossian training plan (2018)). These programs are reflected in more training sessions dedicated to mathematics education. Half the time of inservice training (9 hours) is reserved for mathematics, but this training time is often not handled by mathematics specialists but rather by field educators in charge of these modules. In contrast, the Villani-Torossian training plan (2018), developed from the eponymous report, *21 Measures for Mathematics Education*, is provided by researchers involved in mathematics education with different scientific approaches (didactics, neuroscience, cognitive psychology). This training plan aims to train local mathematical experts over the course of 24 days of training sessions. Their mission is to ongoingly reinforce the support of the teaching teams in mathematics as closely as possible to their field. For example, they will be expected to accompany teachers in schools to design multi-level progressions or to develop teaching sequences on specific mathematical concepts, such as problem solving.

In secondary schools, this training is not mandatory for fully-fledged teachers. In every academic district, several types of training programs are proposed for mathematics teachers: some directly for teaching mathematics (about new technologies, modeling, reasoning, statistics, or even designing spiral teaching), or the link between mathematics and other subjects (history of mathematics, language and mathematics), or about general questions of teaching (e.g., how to manage students with difficulties). Local training, in a lower or upper secondary school, is also recommended to address a specific request or develop teamwork among teachers in a school. As part of the training, mathematics teachers can also prepare other exams (in view of becoming a specialist teacher, trainer, or secondary school headmaster, for example).

The Villani-Torossian training program (2018) promotes a new way of training for secondary school mathematics teachers, using *mathematical laboratories*. Located in an establishment (lower or upper secondary school), these laboratories aim to contribute to teachers' professional development. Although professors (university teachers) can participate in laboratories for the purpose of improving debates, laboratories are essentially places for the school's teachers to discuss, share their practices, and develop cooperation (between the teachers, but also with primary or university-level teachers). These laboratories share certain similarities with Lesson Studies (Quaresma et al., 2018) and are explicitly promoted in the work of Villani and Torossian (2018). Their benefits for professional development have been demonstrated in several countries, but not yet in France.

---

teachers and the CAFFA (*Certificat d'aptitude aux fonctions de formateur académique*) for secondary school teachers.

[21] The 2015 TIMSS (Trends in Mathematics and Science Study) results showed French students lag far behind their counterparts elsewhere in the European Union in both math and science.

Advantages and Disadvantages of Teacher Training Practices in France

The benefits of teacher training in France are multiple, even if disadvantages are perhaps more numerous. The first advantage is the masterization of teacher training, as implemented in 2011. Before this date, teacher training did not require such a high academic level and was not really recognized as academic training. Although it undoubtedly raised the level of teacher training, such masterization also contributed to the incorporation of research in teacher training, and thus, to obliging trainee teachers to be initiated in educational research, and to keep abreast of scientific work in the field of education. The dual training also obligates teachers to consider what happens in the classroom and to gain from the wealth of experience in the field. Trainee teachers face real teaching situations and are, thus, better prepared to teach. They can also use mathematical problems encountered in their classes in their training courses, to be shared and discussed. However, trainee teachers are often afraid to expose their problems (due to class management issues, teaching dilemmas, or professional difficulties) for fear of being judged. Sometimes, educators do not give trainee teachers the opportunity to speak about what happens in their classes. Indeed, training time is so limited and the training program so dense that this is not always possible.

The disadvantages and problems concerning teacher training in France are also severalfold. Some of the most serious include the following. Concerning primary school teacher training in mathematics, the main problem is that it is extremely reduced. How does one train teachers, who generally have a very limited scientific background, significant gaps in mathematics, and sometimes also a bad experience in the subject, to teach mathematics from nursery school (age 3) to grade 5 (age 11) in only 36 hours (in certain ESPEs it could even be less)? Even if mathematics could be introduced in other courses (e.g., analysis of teaching practices or initiation to research), it is not reasonable to think that these training components are sufficient to enable primary school teachers to teach mathematics efficiently and support pupils who struggle with mathematics. Concerning secondary school mathematics teacher training, the problem of training time is not so acute, but for some years now, trainee teachers arrive in ESPE with a low level in mathematics as well as significant gaps.

Another particularity of teacher training in France is the competitive examination that students take to become teachers. These competitive examinations have a significant impact on the training. Indeed, the first year of the MEEF Master's program (for both primary and secondary school), which is half of the initial training, is dedicated to the preparation for the various tests of the competitive examinations. Even if professional tests are supposed to be part of the training, students are more focused on succeeding on the competitive examination than on actually being trained. As all competitive examinations are taken at the end of the first year of the Master's degree, the teacher training program is, in fact, often reduced.

The dual-training mode adopted for teacher training in France raises another issue. This mode of training generates different positions, depending on whether one is an educator or trainee. While trainees are focused on prevailing daily needs, educators are aiming at long-term training objectives. The trainees who are focused on their classes and on gaining tenure at the end of the year have difficulties in projecting themselves into the long term and tend to consider anything they do not use to prepare the next day's lessons to be useless (Gelin et al., 2007). Early in the year, they usually see no benefit in preparing a Master's dissertation during their initial training. In contrast, educators are strongly aware that the initial training is too short and that they have a great training responsibility in the long-term. In their training sessions, educators often try to present content that is useful for a majority of teachers or central subjects that must be taught, but these choices can contrast with trainee expectations. For example, training trainee teachers to teach decimals or fractions during their internship in a nursery school class is a difficult challenge.

CONCLUSION

In this chapter we have attempted to describe teacher training in mathematics in France, distinguishing between primary and secondary school teachers. This training is part of a political and historical context, which is important to know to understand how French teachers in charge of teaching mathematics are trained. However, this description does not reveal local disparities (mathematics professorial level, working conditions, student needs, etc.), which can be stark. Training conditions are not the same in all ESPEs. For example, the number of trainee teachers can vary significantly from one ESPE to another, and this kind of parameter can have significant consequences for the quality of training.

A new reform of teacher training is underway following enactment of the *For a School of Trust* law (July 2019). The ESPE will become an INSPE (Institut National Supérieur du Professorat et d'Education) and the competitive recruitment examinations will move from the end of the 1st year of the MEEF Master's degree to the end of its 2nd year. The content of the MEEF Master's will evolve in the following ways:

For Primary School Teacher Training:

- At least 55% of training time will be dedicated to basic knowledge (reading, writing, counting, respecting others, including knowledge and transmission of republican values);
- At least 20% will be dedicated to versatility (other disciplinary aspects), general pedagogy, and classroom management;
- At least 15% of the time will be dedicated to research;
- 10% of the time will be reserved for context, especially regional, and for innovations specific to each INSPE.

For Secondary School Teacher Training:

- At least 45% of the training time will be dedicated to the discipline and mastery of fundamental knowledge;
- At least 30% will be dedicated to effective teaching and learning strategies, assessment, and classroom management;
- At least 15% of the time will be dedicated to research;
- 10% of the time will be reserved for the specific context and innovations of each INSPE.

These changes are primarily political, and do not suggest that French teachers will be more effectively trained to teach mathematics from nursery school to high school. Measures such as those proposed in the Villani-Torossian training program (the creation of mathematics laboratories in schools, identification and sustainability of clubs related to mathematics) could generate a favorable environment and promote another vision of mathematics teaching and learning, which could contribute to the mathematical professional development of teachers in teams, and ultimately, to the success of all French students in mathematics.

REFERENCES

Brousseau, G. (1998). *Théorie des situations didactiques* [Theory of didactical situations]. La Pensée Sauvage.

Colmant, M. (2015). Teachers, teacher education, and professional development. In I. V. S. Mullis, M. O. Martin, S. Goh, & K. Cotter (Eds.), *TIMSS encyclopedia, France chapter.* http://timssandpirls.bc.edu/timss2015/encyclopedia/countries/france/

Cornu, B. (2015). Teacher education in France: Universitisation and professionalisation: From IUFM's to ESPE's. *Education Inquiry, 6*(3), 289–307.

Duval, A., & Gelamur, A. (2019). *Les mathématiques par les jeux* [Playing and learning mathematics]. Mémoire de Master MEEF Second degré—mathématiques. ESPE de l'Académie de Créteil.

Gelas, C. (2019). *La résolution de problèmes au cycle 2. Evolution des procédures des élèves.* [Problem solving in Cycle 2. Evolution of students' procedures]. Mémoire de Master MEEF Premier degré—mathématiques. ESPE de l'Académie de Créteil.

Gelin, D., Rayou, P., & Ria, L. (2007). *Devenir enseignant. Parcours et formation.* [Becoming a teacher. Courses and training]. Armand Colin.

Gueudet, G., Bueno-Ravel, L., Modeste, S., & Trouche, L. (2017). Curriculum in France: A national frame in transition. In D. R. Thompson, M. A. Huntley, & C. Suurtamm (Eds.), *International perspectives in mathematics curriculum* (pp. 41–69). Information Age Publishing.

Pelay, N. (2011). *Jeu et apprentissages mathématiques : élaboration du concept de contrat didactique et ludique en contexte d'animation scientifique.* [Game and mathematical learning: Elaboration of the concept of didactic and ludic contract in the context of scientific animation]. Doctoral dissertation. Université Claude Bernard - Lyon I.

Quaresma, M., Winsløw, C., Clivaz, S., da Ponte, J.P., Ní Shúilleabháin, A., & Takahashi, A. (2018). *Mathematics lesson study around the world: Theoretical and methodological issues.* ICME-13 Monographs. Springer International Publishing.

Institutional Texts

Arrêté du 1er juillet 2013 relatif au référentiel des compétences professionnelles des métiers du professorat et de l'éducation [Order of 1st July 2013 about the Reference Framework for Professional Skills of the Professions in Teaching and Education]. *Journal Officiel de la République française Paris, 18 Juillet 2013.*

Arrêté du 22 août 2014 fixant les modalités de stage, d'évaluation et de titularisation de certains personnels enseignants et d'éducation de l'enseignement du second degré stagiaires [Order of 22nd August 2014 about the Modalities of Traineeships, Assessment and Tenure of any Education Staff and Secondary School Trainee Teachers]. *Journal Officiel de la République française Paris, 26 aout 2014.*

Décret du 31 mars 2015 relatif au socle commun de connaissances, de compétences et de culture [Decision of 31th March 2015 concerning Common Core of Knowledge, Skills, and Culture]. *Journal Officiel de la République française Paris, 2 avril 2015.* [http://cache.media.education.gouv.fr/file/17/45/6/Socle_commun_de_connaissances,_de_competences_et_de_culture_415456.pdf]

Ministère de l'Éducation Nationale. (2018). *Repères et références statistiques. Enseignement - Formation- Recherche.* [Benchmarks and statistical references. Teaching—Training—Research.] https://cache.media.education.gouv.fr/file/RERS_2018/28/7/depp-2018-RERS-web_1075287.pdf

Ministère de l'Éducation Nationale. (2019a). Devenir enseignant—les programmes des concours d'enseignants du second degré de la session 2020 [Becoming teachers: Program examination for the 2020 session]. http://www.devenirenseignant.gouv.fr/cid98492/programmes-concours-enseignants-session-2020.html

Ministère de l'Éducation Nationale. (2019b). Site du jury du Capes externe de mathématiques [Official Site of Mathematics CAPES's jury]. http://www4.ac-nancy-metz.fr/capesmath/

Ministère de l'Éducation Nationale. (2019c). Site du jury de l'agrégation externe de mathématiques [Official Site of Mathematics Aggregation's jury] https://agreg.org/

Vergnaud, G. (1990). *La théorie des champs conceptuels* [The theory of conceptual fields]. *Recherches en didactique des mathématiques, 10*(2.3), 133–170.

Villani, C., & Torossian, C. (2018). *21 mesures pour l'enseignement des mathématiques.* [21 measures for mathematics teaching]. Paris, France: Ministère de l'Éducation Nationale. http://cache.media.education.gouv.fr/file/Fevrier/19/0/Rapport_Villani_Torossian_21_mesures_pour_enseignement_des_mathematiques_896190.pdf

CHAPTER 4

MATHEMATICS TEACHER EDUCATION IN MALAWI[1]

Everton Lacerda Jacinto[a], Mercy Kazima[b],
Arne Jakobsen[a], and Raymond Bjuland[a]

Over the last decades, there have been numerous plans for addressing issues in teacher education in Malawi. Individual academics and institutions have been creating exciting ways to improve preservice teachers' knowledge and skills to carry out tasks of teaching mathematics more effectively in the classroom. This chapter presents the historical background of the actions taken to qualify preservice teachers for the work of teaching in Malawi. The chapter also gives an overview of the current programs for preparing primary and secondary mathematics teachers and examines the perceptions and some rationales given by mathematics teacher educators about the nature and principles of documents and approaches for preparing preservice teachers in mathematics. The chapter contributes to increasing knowledge about mathematics teacher education in Malawi and its conclusions has implications for future research in this context and in Sub-Saharan countries.

INTRODUCTION

In Africa, teacher education has been an important political instrument for the construction of a democratic system. Despite progress, many of its nations are

[a] University of Stavanger, Norway; [b] University of Malawi, Malawi
[1] This chapter is based on the first author's dissertation (Jacinto, 2020).

International Perspectives on Mathematics Teacher Education, pages 75–92.

struggling to meet the targets set by the United Nations (UN) on Sustainable Development Goal 4 (SDG4) to "ensure inclusive and equitable quality education and promote lifelong learning opportunities for all" (United Nations Educational, Scientific and Cultural Organization [UNESCO], 2016). In the region of Sub-Saharan Africa, for instance, nearly 7 in 10 countries face shortages of teachers (about 6 million teachers) to achieve universal primary and secondary education (UNESCO, 2015a). By 2030, 2.2 million new teaching positions need to be created to achieve the targets, while filling about 3.9 million positions due to attrition (UNESCO, 2015b).

In the context of the Republic of Malawi, which was the first Sub-Saharan African country to implement free primary education, the quality of education has been compromised by the employment of teachers with little or no training (United Nations International Children's Emergency Fund [UNICEF], 2019). Most of the Sub-Saharan African countries are members of the Southern Africa Consortium for Monitoring Educational Quality (SACMEQ), which is a consortium of Ministries of Education in conjunction with UNESCO's International Institute for Educational Planning. The aim of the consortium is to work together to share experiences and apply scientific methods for monitoring and evaluating the quality of education in the participating countries, specifically Botswana, Kenya, Lesotho, Malawi, Mauritius, Mozambique, Namibia, Seychelles, South Africa, Swaziland, Tanzania (Mainland), Tanzania (Zanzibar), Uganda, Zambia, and Zimbabwe. SACMEQ administered tests in numeracy and literacy to primary grade 6 students in all the participating countries. The first test was in 1995, the second in 2000, and the third in 2007. According to the third SACMEQ test, seven in ten children in Malawi lack proficiency in reading and mathematics (SACMEQ, 2010). This situation puts Malawi as one of the least performing countries in Sub-Saharan Africa (Ministry of Education, Science, and Technology, 2016). Looking at the details of performance by Malawian grade 6 students, it was found that more than 90% of the students were operating at a basic numeracy level, which is three grade levels below the expected level of achievement (SACMEQ, 2010).

Kazima et al. (2016) describe many factors that contribute to Malawi students' low achievement, including large class sizes, limited teaching and learning resources, and the poor quality of teachers. Although these factors are connected and need to be addressed together to improve the quality of education in Malawi, the quality of teachers is most important because a well-qualified teacher will be able to cope and teach better within the limited circumstances of the Malawi context than an unqualified teacher (Kazima, 2014). Furthermore, the other factors also affect countries like Lesotho and Zimbabwe, whose students performed better than students in Malawi. Hence, the quality of teaching seems to be the main factor for Malawi's students' low achievement. This is likely to be a consequence of employing unqualified teachers to meet the high demand for teachers that was created by the introduction of free primary education.

This chapter is based on research that explored the reality of teacher education in Malawi, particularly the primary school sector. The goal of this study was to comprehend better the specificities, needs, and challenges faced by teacher educators in the context of Malawi. The chapter provides an overview of current requirements to become primary and secondary teachers in Malawi, as well as possibilities and limits of expansion of pedagogical practices in these areas.

BACKGROUND: THE REPUBLIC OF MALAWI

Malawi is a landlocked southern African country that became a republic in 1966 following its independence from Britain in 1964. Prior to independence, Malawi was called Nyasaland and was a British protectorate (Chanaiwa, 1993). From independence in 1964 to 1994, there was only one party, and leadership was not democratic (Kazima, 2014). In 1994, other parties were introduced, the first presidential and multiparty elections were held, and a new party came into power. Since then, Malawi has been a multiparty democratic nation, and so far, three different parties have been in power. Malawi is densely populated. It has an area of approximately 118,500 square kilometres and a population of about 18.7 million people (Government of Malawi, 2018). Nearly one-tenth of the population lives in urban areas (11.4%), whereas the remaining majority live in rural areas (88.6%) (Yaya et al., 2016). Malawi is also one of the poorest countries in the world, ranked as 171 out of 189 countries on the Human Development Index (United Nations Development Program, 2018).

Malawi has struggled to provide quality education in schools due to the large population and the limited resources available. Consequently, in past years, Malawi has become a focus of concern for world leaders due to the poor indices in reading and mathematics competencies in schools (SACMEQ, 2010). With the introduction of free primary education in 1994, Malawi now faces multidimensional challenges, such as lack of facilities, high pupil/teacher and pupil/classroom ratios, low learning achievement, and poor qualification of teachers. In urban districts, the average number of pupils per teacher is about 60. In contrast, in rural areas of the country, for instance in the Mangochi district, the rate reaches 1 teacher per 152 pupils (UNICEF, 2019). Over the years, the government has introduced some interventions in an effort to address the problems, for example, double shifts in schools to address the problem of classroom space and hiring many unqualified teachers to meet the high demand for teachers (UNICEF, 2019).

MALAWI EDUCATION SYSTEM

The Malawi education system is divided into three levels: primary, secondary, and tertiary (postsecondary) education. In primary school education, children's ages range from 6 to 14 years old, ascending from *Standard 1* to *Standard 8*. Secondary school education is offered for students with ages between 15 to 19 years old, occurring from *Form 1* to *Form 4*. However, due to students repeating a grade and other reasons, the age range might exceed 14 and 19 for primary and secondary school, respectively. The duration of tertiary education varies depending on the programs and institutions. For example, programs at a university often take four to five years, and vocational education can take one to two years. Pre-primary education in Malawi is a non-mandatory program designed for children ages 3 to 5 and offered by *Early Childhood Development Centres* that are managed by the Ministry of Gender, Women, and Child Welfare.

TEACHER EDUCATION IN MALAWI

In this section, we first discuss the program for primary teacher education. Then, we discuss three different pathways for secondary teacher education.

Primary Teacher Education

Primary teacher education in Malawi started during the colonial period. Primary teacher colleges were established by churches in association with the Christian Council (Banda, 1982). All the teacher colleges followed a three-year program with the same curriculum. The three-year curriculum contained more content courses than the current curriculum. After independence, the Ministry of Education took over responsibility and management of primary teacher education and established more teacher-training colleges (TTC). As of 2019, eight teacher-training colleges exist.

Since the introduction of free primary education in 1994, there has been a rapid increase in the number of children in primary schools—from 1.8 million in 1993 to 4.5 million in 2014 (Government of Malawi, 2016). This resulted in a very high demand for teachers and the Ministry employed about 22,000 unqualified teachers with a plan of offering them on-the-job training. The minimum requirement for the recruitment was a Junior Certificate, which was obtained after passing national examinations at the end of two years of secondary education. Therefore, these teachers had only two years of secondary school mathematics, which include Algebra, Euclidean Geometry, and Arithmetic. Consequently, the teachers' own knowledge of mathematics was not much higher than the primary school mathematics they were required to teach.

As another intervention to address the problem of the shortage of qualified teachers, the Ministry reduced the duration of the teacher education program from three years to two years (Kazima, 2014). After recruiting thousands of unqualified teachers, the Ministry discontinued the full-time program for teacher education and introduced a largely school-based program in 1997, called the Malawi Integrated In-Service Teacher Education Program (MIITEP). The program took twenty-four months, with four months for college-based courses and the remaining 20 months for teaching in schools. The college-based courses included Educational Foundations, English, Mathematics, Science, and Expressive Arts. The mathematics included mostly Arithmetic and mirrored the topics of primary school mathematics. The school-based learning was in the form of teaching practice with supervision by the college lecturers and the classroom teachers. The college lecturers were recruited by the Ministry of Education and were required to have more than 10 years of teaching experience in primary schools and minimum qualification of Diploma in Education, which is lower than Bachelor's degree level.

The MIITEP was supposed to be an emergency program for about five years. However, it continued to run for ten years until 2005 because more unqualified teachers continued to be employed. In 2005, the Ministry discontinued the MIITEP and introduced a new program called the Initial Primary Teacher Education Program (Ministry of Education, 2005).

The new Initial Primary Teacher Education Program had a duration of two years. The first year consisted of full-time college-based courses; the second year was full-time school based in the form of teaching practice. The entry requirement into the program was also revised from the Junior Certificate to the Malawi Schools Certificate, which requires passing a national examination after completing the four years of secondary education. The examinations are equivalent to O-Level in the English system. The candidates for this certificate would have covered Algebra, including algebraic expressions, linear equations, quadratic equations, graphs of equations, and inequalities; Euclidean Geometry, including lines and angles, polygons and their properties, circle properties and theorems; Arithmetic, specifically commercial arithmetic; and Trigonometry, including sine, cosine, tangent, and trigonometric identities. The curriculum of the Initial Teacher Education Programme had a total of ten learning areas: agriculture, education foundation studies, environmental studies, expressive arts, general and social studies, life skills, literacy and languages (English and Chichewa[2]), numeracy and mathematics, religious studies, and science and technology. All preservice teachers were required to study all ten learning areas because primary school teachers are expected to teach all the subjects in primary school (Ministry of Education, 2005). This curriculum was offered from 2005 to 2016.

In 2017, the curriculum was revised, and the structure of the program was changed from one year (three terms) of college courses and three terms of teaching practice to four terms of college courses and two terms of teaching practice. In this new structure, preservice teachers take theoretical courses in college for the first two out of three terms of the first year, learning subject content with a special focus on teaching methods for primary classes. After completing the coursework in these two terms, preservice teachers go into the field for practice, experiencing teaching in both lower and upper primary classes for two terms (last term in first year, and first term in year two). During teaching practice, the students are supervised periodically by the teacher educators and more regularly by mentors in the school. Mentors are experienced teachers who have been trained by the teacher colleges to supervise student teachers during teaching practice in the schools. In

---

[2] English is the official language of Malawi while Chichewa is the national language.

the last two terms of year two, the preservice teachers return to college to continue learning subject content, reflecting on their experience of teaching practice, and then wind up their studies (Malawian Institute of Education, 2017).

The current revised curriculum has adopted a reflective practitioner model of teacher education (Schön, 1987) that aims to connect practice and theory, and integrates content and pedagogy in teaching and learning. The innovative features of this curriculum include:

- The curriculum design is based on reflective practice principles;
- Introduction of specific early grade teaching methodologies;
- Delivery of the subject content following modular approaches;
- Teaching experience in both lower classes and upper classes of primary school; and
- Cross-cutting studies involving assessment for learning, information and communication technology, and critical thinking.

The primary teacher education curriculum is designed in a modular structure and contains the same ten learning areas outlined earlier in this chapter, but with different weighting in terms of time. Three learning areas (Education Foundations Studies; Numeracy and Mathematics; and Literacy and Languages) are allocated more time than the rest. In this modular design, a set of topics forms a module in a subject. A module consists of 40 hours of contact time. The core elements in Numeracy and Mathematics are shown in Figure 4.1.

Primary teacher education programs have both continuous (formative) and summative assessment to measure preservice teachers' achievement of knowledge, skills, values, and attitudes. These take the form of oral presentations, practical tasks, reports, research, tests, and examinations. The grading system is given in percentages: 60% for continuous assessment and 40% for summative assessment.

The current mathematics curriculum for primary teacher education encompasses standards that require preservice teachers to acquire subject matter knowledge and to develop pedagogical ways to promote students' lifelong learning (Malawian Institute of Education, 2017). The curriculum contains three major goals for teachers in the teacher education program:

1. Develop academically well-grounded and professionally competent teachers;
2. Stimulate flexibility and capability of adapting to the changing needs and environment of Malawi society; and
3. Professionalism in adhering to and maintaining the ethics of the teaching profession and being imaginative in adapting, creating, and utilizing locally available resources suitable for the needs of their learners. (Malawian Institute of Education, 2017, p. ix)

Core Element	Term 1	Term 2	Terms 3 and 4	Term 5	Term 6
Theories, Concepts, and Issues in the Teaching and Learning of Mathematics	40				
Number Concepts and Operations (e.g., whole numbers, fractions, decimals, percentages, rate, ratio, proportion)	40	40			
Measurement (e.g., length, perimeter, area, volume)		40	Teaching Practice		
Data Handling (e.g., representing data, bar graphs, pictographs)				40	
Space and Shapes (e.g., triangles, rectangles, circles)				40	
Accounting and Business Studies (e.g., profit and loss, simple and compound interest)					40
Patterns, Functions, and Algebra (e.g., number patterns, linear equations)					40

FIGURE 4.1.   Core Elements in Numeracy and Mathematics by Term and Number of Hours (Malawian Institute of Education, 2017)

The curriculum rests its philosophy on the basis of the following principle: "To produce a reflective, autonomous, lifelong learning teacher, able to display moral values and embrace learners' diversity" (Malawian Institute of Education, 2017, p. ix). Therefore, preparing teachers of quality seems to be an important goal within this context, but still it is unclear how these principles and the main principles guide teachers' practice. The curriculum for preparing primary teachers in Malawi focuses on three main skills needed for an effective teacher: competency, flexibility, and resourcefulness. However, no transparency on the meaning of these skills is provided, for instance, what is meant by "being a resourceful teacher in Malawi"?

Secondary Teacher Education in Malawi

There are three routes of study to becoming a secondary school mathematics teacher in Malawi: (i) Bachelor of Education, (ii) Diploma in Education, and (iii) Bachelor of Science followed by University Certificate of Education. We discuss each of these in the following sections.

Bachelor of Education

The Bachelor of Education program is offered by universities. In Malawi, for a long time there was only one university, the University of Malawi, which is a public university. The University started in 1968 as a College of Education and then was upgraded to a university in 1973 when it included other programs besides education. From then to 1999, the University of Malawi was the only institution that had a teacher education program for secondary school teachers. In 1999, a second public university was opened, Mzuzu University (Saiti et al., 2014). To date, these are the only two institutions for secondary teacher education with Bachelor qualification. The education program can be in one of three specifications: Science Education, Language Education, and Social Studies Education. Mathematics falls under Science Education. We discuss in more detail below the structure for the University of Malawi mathematics teacher education program. The structure for the Mzuzu University program for mathematics teacher education is fairly similar.

Entry into the University of Malawi's Bachelor of Education program requires passing the Malawi School Certificate of Education (MSCE) examinations, which are national examinations at the end of four years of secondary school. For the Education Science program, candidates are required to have passed with high grades in Mathematics, Biology, Physical Science, and English. By the time of the examinations, candidates would have studied four years of mathematics that includes Algebra, Euclidean Geometry, Arithmetic, Trigonometry, and Elementary Statistics. There are many more candidates who qualify for the science program than there are places available at the University. Therefore, entry is very competitive with only about the top 10% of qualifying candidates being accepted into the program (University of Malawi, 2017). The design of the program is that students study mathematics content courses and other science content courses from the Faculty of Science, and the students study education courses from the Faculty of Education. The courses from the Faculty of Education include Mathematics Education and Educational Foundations. The entire program takes 4.5 years and the structure is as shown in Figure 4.2.

All students admitted by the Faculty of Education are taught mathematics courses together with students from the Faculty of Science. By the end of the program, the mathematics preservice teachers have as much content in their mathematics courses as Bachelor of Science students.

Diploma in Education

The Diploma in Education is a three-year program offered by Colleges of Education, which is administered by the Ministry of Education. There are currently two colleges offering this program, and both are public colleges. The first, Domasi College of Education, was opened in 1993; the second, Nalikule College, was opened in 2017. The two colleges have the same curriculum and are both affiliated with the University of Malawi. The curriculum they offer includes mathematics content courses, mathematics education courses, education foundations courses, and teaching practice. In comparison to the University of Malawi's Bachelor of Education program, the colleges overall offer content up to about the third-year level of the Bachelor of Education program and their graduates have about half the level of mathematics the Bachelor graduates have. The two colleges were

Year	Courses (total of five courses each year)
1	One course on English language communication skills from the Faculty of Humanities Four courses from the Faculty of Science: Mathematics (College Algebra, Trigonometry and Calculus) and three others (Biology, Chemistry, Physics, Earth Science, or Home Economics) No courses from the Faculty of Education
2	Two Mathematics courses (Calculus, Linear Algebra, Discrete Mathematics, Financial Mathematics, or Mathematical Computing) and two other science courses from the Faculty of Science One Educational Foundations course (Educational Psychology or Sociology of Education) from the Faculty of Education
3 and 4	Three Mathematics courses per year (Calculus, Real Analysis, Complex Analysis, Abstract Algebra, Numerical Analysis, Differential Equations, Graph Theory, or Number Theory) from the Faculty of Science One Mathematics Education course per year (Mathematics for Teachers, Mathematics Teaching Studies, or Curriculum Studies in Mathematics) from the Faculty of Education One Educational Foundations course per year (Philosophy for Teachers, Special Needs Education, Educational Technology, History of Educational Thought, Economics of Education, Sociology of Education, or Psychology of Education) from the Faculty of Education
5	Teaching practice in schools for 10–12 weeks

FIGURE 4.2.   Structure of the Science Program for Mathematics at the University of Malawi (University of Malawi, 2009)

opened by the government to increase the number of secondary school teachers, and also to produce teachers at a faster pace than the universities (Saiti et al., 2014).

Bachelor of Science followed by University Certificate of Education (UCE)

The Bachelor of Science followed by University Certificate of Education route is for teachers who have the subject qualification but not the teaching qualification. For mathematics teachers, they would already have a Bachelor of Science with mathematics as a major. Such teachers are often employed by the government because of the shortage of mathematics teachers, but they are considered unqualified in terms of teacher qualification. Therefore, they are expected to study the University Certificate of Education (UCE) course that qualifies them as teachers. This is a part-time course offered during school holidays and the unqualified teachers take the course while on the job. It is also possible to study the UCE course before seeking employment as a teacher.

Other Options

Finally, it is important to acknowledge that within the last decade, some private universities have opened in Malawi and most of these also offer teacher education programs. However, the number of students who attend these is relatively small compared to the public institutions.

PRINCIPLES FOR PREPARING PRIMARY PRESERVICE TEACHERS: CURRICULAR VIEWS FROM MATHEMATICS TEACHER EDUCATORS

The existing teacher training guidelines define, to some extent, the characteristics needed by a teacher to attend to the demands of society. In the context of Malawi, one specific characteristic of an effective teacher is that of being *resourceful*. Such a component provides significant contributions to both primary and secondary teacher education in Malawi.

In the following sections, there is an analysis of the views of two Malawian teacher educators from primary teacher education. Their views were obtained via individual interviews in a Teacher Training College (TTC). In this section we explore, from the perspective of teacher educators in mathematics, the profile of a resourceful teacher, as well as the needs and challenges for the preparation of preservice teachers in Malawi. These insights are crucial to understanding the nature and role that institutional documents play in shaping teacher education in practice. This section is divided into three categories: i) The meaning/value that resourceful teaching has on the preparation of preservice teachers in Malawi; ii) The relationship between the specificity of being resourceful and the principles of the new curriculum; and iii) Needs and challenges for preparing primary teachers in mathematics.

What is the Meaning/Value that Resourceful Teaching Has on the Preparation of Preservice Teachers in Malawi?

Guidelines for training primary preservice teachers in mathematics emphasize producing "A reflective, autonomous, lifelong learning teacher, able to display moral values and embrace learners' diversity" (Malawian Institute of Education, 2017, p. ix). To achieve this goal, primary teacher training colleges value the quality of being resourceful. It seems to be an important characteristic needed for preservice teachers by the end of their TTC program as it is of particular importance for the context of Malawi. The idea of resourcefulness appears in the curriculum as a way to overcome the scarcity of didactical materials and circumstances for teachers and preservice teachers to facilitate the learning of students. This is important because it helps institutions and the government not only to rethink the educational policies but also to rethink pedagogical actions for preparing preservice teachers.

For Anita (pseudonym), a primary school teacher educator in mathematics who has been working several years in a teacher training college in Malawi, the concept of resourcefulness is related to the use of locally available resources; that is, the ability of teachers to make use of and transform resources they find in their local environment into didactical means for teaching and learning:

> In our primary schools, we do not have materials that you can use for teaching. As it happens in most of the developed countries, schools are supposed to provide materials for teachers to use, right? But, in Malawi, schools do not have the capacity to provide resources to the teachers. So, if a school does not provide resources to the teacher, should the learners not learn using resources? No! They should learn, and the teacher should take the initiative to make resources themselves. That is why if you go to some classes in Malawi, you will see that the preservice teacher has made place-value boxes[3] for the learners, so they can have more opportunities to explore the content. (Anita, Interview)

A second mathematics teacher educator provided a similar view as that of Anita. John (pseudonym), a primary teacher educator in mathematics, explained that if schools in Malawi cannot afford to buy didactical materials, teachers should take the initiative to produce these materials themselves. John expresses this in the following way:

> Resourceful means someone who does not wait for the government to provide the resources or stop to teach because there is no material to teach. The meaning of being resourceful is someone who is able to find different ways of teaching, different resources that help them to teach more effectively. We are not talking about only materials. They [teachers] can go to another school to learn with them; they can bring their students to visit places to meet people. So you can borrow these things and use them to teach. That means being resourceful! (John, Interview)

As articulated by John, teachers need to be proactive regarding not only the use of materials, but they need knowledge to anticipate what they will encounter during lessons and take the proper initiative rather than waiting for something to occur. A resourceful teacher who develops the ability to satisfy students' learning needs and specificities seems to be consistent with what effective teaching means for John. As part of this conception, John believes that a teacher should focus on classroom diversity regardless of institutional and/or governmental support. By acknowledging and using different strategies in the classroom, the teacher becomes resourceful and capable of enhancing students' learning with situations within his or her own realities. Teaching should not only satisfy instructional demands or promote students' learning but should also exhibit a variety of styles and incorporate real-world applications.

---

[3] Place-value box is a tabular resource used for teaching and learning of numeracy. In Malawi, teachers produce place-value boxes from locally available resources such as paper and cardboard. It consists of three compartments containing pieces of paper, sticks, stones, bottle tops, and seeds that are used to represent ones, tens, and hundreds. Normally, these boxes are used for primary school teachers in mathematical lessons about place value.

Quality and shortage of resources is also a concern for John. Although accessing didactical materials is a factor that contributes to effective teaching, he explained that it should not be seen as a factor to measure its quality:

> We should not devaluate the quality of a teacher because they don't have access to resources. We want to prepare them to find these resources on their own, so they have to be creative now [on] their own and think: "How can I find this resource? Or better, how can I produce something that will work with these resources that I have?" At TTC, for example, we can provide them with a marker, but when they go to schools, they cannot find it there. But, still, they need to have something [with which] to write. So what do they do? They can get some pieces of charcoal and use it to write. So, we prepare them for that kind of situation. Another is, if they want to produce green cards, they can get some leaves from the trees and use them as a green card. So, this is something that is locally available from nature. (John, Interview)

In the excerpt above, we see that resourcefulness is not the only component that preservice teachers need to achieve in their training. Also, skills of improvisation, classroom management, and critical thinking that help them to grow independently should be considered. Although the literature suggests that teaching materials are an essential component for refining teachers' quality (Allwright, 1981), John also said that the knowledge of the purposes and goals of using a particular resource should be added in the preparation of preservice teachers. In teacher training, preservice teachers are required not only to develop their perception of how to utilize locally available material, but also their creativity for finding different means to disseminate ideas or knowledge in classrooms. Therefore, the quality of being resourceful, as described by John, seems to serve as a channel for teachers through which they can assist students' leaning, by transforming objects and natural resources into pedagogical and instructional materials.

How Does This Idea of Preparing Preservice Teachers to Become Resourceful Teachers Fit the Philosophy of the TTCs?

John and Anita were asked to provide insights on the characteristics that the new curriculum suggests be developed in preservice teachers. The first characteristic is reflectivity. According to John:

> Every teacher is supposed to be reflective! After you conduct a lesson, you ask: "How did I use it? How did the learners get the staffs [content] that I was teaching? Ah… Instead of this method, I could have used this one!" So, thinking in this way, you are reflecting on what you have done. So, the next time you are in the class, you can come up with a different approach and use it. It is different from our friends in Japan that are using lesson study. In lesson study, you are able to reflect as a group. So, you teach, they observe you, and then you reflect together. You and the other teachers reflect together as a group. But here in Malawi, you have to reflect as an individual. There are not many teachers that you can talk [to] about the classes. This

is an issue we have here. So, for us, reflective means being an individual, reflective [person]. (John, Interview)

Anita also corroborated with a similar view:

About a reflective teacher, he needs to think: "How is my teaching going? The resource[s] that I use did they work well? Have they helped my learners to learn? Or, maybe, how should I improve them? Or, maybe next time, should I use different resources? Or, this time, did I go to class with resources that I am supposed to? But next time, should I use resources?" That is what we expect them to think. The teacher needs to reflect on his teaching, in the materials he is using, in the preparation of his lesson. (Anita, Interview)

We can see that for both John and Anita, reflection in teacher education involves experience that one has already done or any action that has been taken, and what could have been done differently. By reflecting on practical experiences in the classroom, teachers tend to develop critical thinking of their own ideas and actively search for new means of dealing with complicated situations (McKnight, 2002). This type of mental skill is crucial for teachers, especially for beginning teachers, to develop while they are preparing for their profession (Brookfield, 1995; Clarke et al., 2013). Reflective thinking helps one to promote an awareness of cognitive habits in order to shape his or her own (Ryan & Cooper, 2006). Moreover, reflective thinking helps preservice teachers increase both cognitive flexibility and effectiveness when learning theories and addressing practical problems in the classroom. It can be useful for understanding how different approaches might promote students' curiosity, joint problem solving, and reflection on the perspective of others (Berk & Winsler, 1995).

Both Anita and John provided full accounts of teacher autonomy. This feature appeared to encompass personality and disposition to tackle the tasks of teaching. John describes how the idea of autonomy in the preparation of teacher education has changed over time:

In the past, we have been relying on everything in the books: activities, examples, exercises... so you can find it in the book. But today we are saying no! This is not the type of teacher we want. We want a teacher who is able to say: "Right, this is my topic; how can I get it interesting for the students, but at the same time how can I cover what is proposed in the syllabus?" This is what we are looking for. Teachers who are able to construct a lesson on his or her own way to fit the learners' needs, so that they can become autonomous... they can expand what they know, which means they will not follow everything from what is written. (John, Interview)

Anita described this pattern of thinking in the following way:

If we are talking about an autonomous teacher, for example, what does it means? It means a self-[starter], somebody who can solve a problem by himself. As we have a problem with resources, we need a teacher able to ask himself: "I need resources for teaching this subject, but the school does not provide, what should I do?" So he or she needs to pay attention to this and find a way to make the resources. (Anita, Interview)

Resourcefulness had less emphasis in John's views than it did for Anita. On the one hand, John spoke about the need for training preservice teachers to be able to balance both students' needs and curriculum demands. Anita, on the other hand, referred to the resourcefulness as a basis for the development of preservice teachers' autonomy. She referred to a proactive attitude of inquiring and reflecting on the ways to solve the problem with a lack of resources. Among the enduring individual preservice teachers' characteristics, which she perceived as valuable, were abilities (general and specific), such as predisposition, motivation, and attentiveness. This suggests that Anita placed emphasis on what resourcefulness means for the maturity of teachers' autonomy, whereas John placed emphasis on how the quality of resourcefulness can contribute to the organization of the lesson and the development of autonomy.

A third way in which resourcefulness relates to general principles of the teacher curriculum arises from the preservice teachers' capacity of positioning themselves as learners:

What about a lifelong learning teacher? There is this new development of our educational system, so we need teachers who can use locally available resources. So he or she should be able to understand and first to learn it, he or she needs to learn every day how to do it. So, as teachers, we are always learning how to come up with our own resources for teaching. Thus, he can make the resources by learning how to find, create, and select the right materials. (Anita, Interview)

In contrast, John demonstrated a general perspective of preservice teachers' learning that continues into their future careers:

As a teacher, we should keep learning. Because teaching is also about learning, right? For every meeting, every class, every moment with the students, at least we should learn something. If we go to CPD [Continuing Professional Development], we should at least come up with something. That is why we should keep learning. For example, as I said, in Malawi we don't have lessons. But people can say this… and this… it is very important; and we need to learn this. So we take that, and think how to implement it here. That is how we keep learning as a teacher. We should not reach a point that I have learned a lot, and that is it. So, preservice teachers need to understand this when they are preparing to become a teacher. They also need to see them[selves] as learners. (John, Interview)

Both teacher educators also saw students' diversity as an important component of their needing to become resourceful. One aspect of students' diversity that they mentioned was students with special needs. Anita explained how students' needs influenced her teaching, specifically in the decisions she made about material resources:

In terms of student diversity, for example, we got a class where there is a learner who is blind. So, the image you are going to bring to the class with resources is important but it will only work with those who are able to see. But what about the learner who is blind? How are you going to embrace diversity in that case? You have to think and make resources that can help this particular learner to learn. I don't know if the developers of the curriculum had this same idea. But, that is how I see. It works as a big umbrella and each component has to fit here. But the way ... I look at it, as a teacher educator, that is how I understand. That is what I think how this idea of being resourceful fits the idea of student diversity within this new curriculum. But, of course, I cannot answer on the part of the curriculum developers. (Anita, Interview)

John showed an appreciation for the mediational role as material conditions interact with students' diversity for learning the content:

Yes, we emphasi[ze] resourcefulness. But, this is very important for them to explain abstract concepts. They need to make sure that every student has the opportunity to visualize the concepts... that they embrace students' diversity. More specially, when we are in infant classes, so they have problems to understand abstract concepts. (John, Interview)

Mastering particular content was not of major significance for Anita. When asked how resourcefulness might be related to the principle of displaying moral values, she mentioned the role a teacher plays in the classroom:

The teacher should be a model for the learner. At TTC, we try to teach them how the learners can encourage someone who provided a wrong answer for a mathematics problem, so the teacher needs to encourage him or her. It is not about saying, "This answer is an incorrect answer," but the teacher needs to consider what the learner has done. That is why in many class[es], after someone comes to the blackboard and solve[s] a problem, the teacher asks the students to congratulate them for the effort made. (Anita, Interview)

While describing that classrooms in Malawi are crowded and difficult environments for teachers to provide quality teaching, John made reference to the need for using didactical strategies that might help teachers to display moral values to the students:

Able to display moral values can start right there before [learning] to [be] resourceful. The teacher should look up what students are doing. So, there are some criteria we expect them to follow. For example, as we have many children in the classroom, you can create small groups. So, it will be easier for the teacher to assess them and check if everyone is participating and making a contribution. (John, Interview)

Needs and Challenges for Preparing Primary Teachers in Mathematics

The revised curriculum for preparing primary teachers in Malawi has embraced key elements that reflect current thinking on effective teaching, a thinking that is built upon the improvement of teachers' skills, school facilities, and students' performance in different subjects. Although, in theory, such principles seem to be a reliable way to cover urgent gaps in the educational context in Malawi, in practice, they reveal a shortcoming in the preparation of future teachers. For teacher educator Anita, one of the main issues in preparing preservice teachers for the work of teaching in primary schools is the lack of devotion to the duty for becoming teachers:

> There are requirements that they have to satisfy for them to be accepted into teaching. The applicants need to have a secondary school certificate (MSCE) with six credits, one of which is a credit in English and any other science subject. After applying, they are invited for interviews. When they are successful, they are selected to attend the TTC. There are some who choose to join teaching because they want to be teachers, but there are others who apply only because they have not been picked in other professional programs, so they have no choice but to find something to do. And this is a big problem! (Anita, Interview)

In terms of needs and expectations for the preparation of preservice teachers in mathematics, teacher educator John explained some of the complications for improving the quality for preparing teachers within the educational system:

> The first challenge is the system we are using to take our teachers to college. If we look at the advertisements, they are saying that the one to qualify as a teaching profession should have English, any sense of a subject that can catch up such as biology and physics of science. Any other two subjects should have a credit on that. But there is no mention of mathematics. Meaning, students don't need to have mathematics to come here, so we can have people prepare to become teachers who failed at mathematics. So, that is a big challenge here, because we don't focus much on the content, but more on the pedagogical content knowledge. (John, Interview)

In terms of challenges in the preparation of preservice teachers in mathematics, teacher educator John explained that the problem comes from the system itself. According to John, a lack of mathematics knowledge for preservice teachers is seen as a "big problem" for student candidates coming to teacher training programs. The academic principles of teacher training programs in Malawi focus more on the development of pedagogical skills for teaching. This could facilitate the work of teacher educators once the new curriculum assumes that mathematics skills are increasingly necessary for the work of elementary teachers in Malawi (Malawian Institute of Education, 2017).

CONCLUDING REMARKS

Teacher education has been an essential instrument in building a democratic system in Malawi. Recent reforms in the curriculum for preparing both primary and secondary preservice teachers in mathematics have aimed for inclusive and equitable quality education and lifelong learning. These qualities encompass characteristics that are both unique and globally shared (profound knowledge of the subject matter and pedagogical content, abilities to work with larger classrooms and poor environmental conditions, and cultivation of specialized knowledge of resourcefulness). By exploring views and insights from two teacher educators from a teacher training college in Malawi, we delve into how they understand the particular principle of resourcefulness that underpins the new primary teacher training curriculum (Malawian Institute of Education, 2017). In this context, we explored how teacher educators understand not only tasks of teaching in mathematics, but also what knowledge and abilities fit the basis of preservice teachers' formation during the teacher-training program. This information provides a better understanding of how the educational system for preparing elementary teachers in Malawi is constituted, providing implications to produce guidelines and policies in teacher education.

REFERENCES

Allwright, R. L. (1981). What do we want teaching materials for? *English Language Teaching Journal, 36*(1), 5–18.

Banda, K. N. (1982). *A brief history of education in Malawi*. Blantyre, Dzuka Publishing Co.

Berk, L., & Winsler, A. (1995). Scaffolding children's learning: Vygotsky and early childhood education. *NAEYC Research into Practice Series, Vol. 7*. National Association for the Education of Young Children.

Brookfield, S. (1995). *Becoming a critically reflective teacher*. Jossey-Bass.

Chanaiwa, D. (1993). Southern Africa since 1945. In A. Mazril & C. Wondji (Eds.), *General history of Africa VIII: Africa since 1935* (pp. 249–281). Heinemann UNESCO.

Clarke, D., Hollingsworth, H., & Gorur, R. (2013). Facilitating reflection and action: The possible contribution of video to mathematics teacher education. *Journal of Education, 1*(3), 94–121.

Government of Malawi. (2016). *Education sector performance report 2015–16*. Ministry of Education, Science, and Technology.

Government of Malawi. (2018). *2018 Malawi population and housing census: Main report*. National Statistical Office.

Jacinto, E. L. (2020). *The development of pre-service teachers' understanding of the knowledge necessary to teach mathematics: A case study in Malawi*. University of Stavanger, Stavanger, Norway. (UIS:555).

Kazima, M. (2014). Universal basic education and the provision of quality mathematics in Southern Africa. *International Journal of Science and Mathematics Education, 12*(4), 841–858.

Kazima, M., Jakobsen, A., & Kasoka, D. N. (2016). Use of mathematical tasks of teaching and the corresponding LMT measures in Malawi context. *The Mathematics Enthusiast, 13*(1–2), 171–186.

Malawian Institute of Education. (2017). *Syllabus for initial primary teacher education: Mathematics ministry of education, science and technology*. Malawi Institute of Education.

McKnight, D. (2002). *Field experience handbook: A guide for the teacher intern and mentor teacher.* University of Maryland.

Ministry of Education. (2005). *Initial primary teacher education teaching syllabus: Numeracy and mathematics.* Malawian Institute of Education.

Ministry of Education, Science, and Technology. (2016). *Malawi government education statistics.* Department of Education Planning.

Ryan, K., & Cooper, J. M. (2006). *Those who can, teach.* Houghton Mifflin.

Saiti, A., Kyle Jr., W. C., Sinnes, A. T., Nampota, D. C., & Kazima, M. (2014). Developing relevant environmental education in a rural community in Malawi. *The Brazilian Journal of Research in Science Education (RBPEC), 14*(2), 185–198.

Schön, D. A. (1987). *Educating the reflective practitioner: Toward a new design for teaching and learning in the professions.* Jossey-Bass Higher Education Series (SAC-MEQ).

Southern and Eastern Africa Consortium for Monitoring Educational Quality. (2010). *SACMEQ III project results: Pupil achievement levels in reading and mathematics.* http://www.sacmeq.org

United Nations Development Program. (2018). *Human development indicators and indices: 2018 statistical update team.* http://hdr.undp.org/sites/default/files/2018_human_development_statistical_update.pdf

United Nations Educational, Scientific and Cultural Organization (UNESCO). (2015a). *Regional overview: Sub-Saharan Africa.* Education for all Global Monitoring Report 2015. https://en.unesco.org/gem-report/sites/gem-report/files/regional_overview_SSA_en.pdf

United Nations Educational, Scientific and Cultural Organization (UNESCO). (2015b). Sustainable Development Goal for Education cannot advance without more teachers. *UNESCO Institute for Statistics Fact Sheet 33.* https://unesdoc.unesco.org/ark:/48223/pf0000234710

United Nations Educational, Scientific and Cultural Organization (UNESCO). (2016). *Unpacking sustainable development goal 4 education 2030 guide.* http://unesdoc.unesco.org/images/0024/002463/246300E.pdf

United Nations International Children's Emergency Fund (UNICEF). (2019). *Education budget brief: Towards improved education for all in Malawi.* https://www.unicef.org/esa/sites/unicef.org.esa/files/2019-04/UNICEF-Malawi-2018-Education-Budget-Brief.pdf

University of Malawi. (2009). *University of Malawi calendar.* Central Office.

University of Malawi. (2017). *Bachelor of Education mathematical sciences programme document.* Chancellor College.

Yaya, S., Bishwajit, G., & Shah, V. (2016). Wealth, education and urban–rural inequality and maternal healthcare service usage in Malawi. *BMJ Global Health, 1,* e000085.

CHAPTER 5

MATHEMATICS TEACHER EDUCATION IN SINGAPORE

Eng Guan Tay and Berinderjeet Kaur

National Institute of Education, Nanyang Technological University

The National Institute of Education (NIE), an institute of the Nanyang Technological University (NTU), is the sole institute for teacher education in Singapore. Unlike many other countries, in Singapore the Ministry of Education (MOE) has sole responsibility over the recruitment, preparation, certification, appointment, and deployment of teachers for schools. The MOE recruits teachers from the top one-third of each cohort of the graduating class who qualify for tertiary education; only one of eight applicants interviewed is accepted. At the NIE, there are two programmes for preservice mathematics teachers, an undergraduate programme and a post-graduate programme. In both programmes, active research mathematicians work together with mathematics educators to prepare future mathematics teachers, thereby ensuring that their mathematics knowledge and pedagogical content knowledge are robust. Systematic on-going review and revision of content and pedagogy courses for the programmes ensure that mathematics teacher education is purposeful and abreast of international trends.

BACKGROUND

In Singapore, education for primary, secondary, and tertiary levels is generally supported by the national government. All institutions, private and public, must

International Perspectives on Mathematics Teacher Education, pages 93–111.

be registered with the Ministry of Education (MOE). English is the language of instruction in all public schools and tertiary institutions. At all grade levels, all subjects, including English as a first language, are taught and examined in English except for the *Mother Tongue* language and other foreign languages, such as Arabic, French, and German that students may opt to do as a third language in their secondary school. Although *Mother Tongue* generally refers to the first language internationally, in Singapore's education system it refers to the second language (which is usually the language of one's ethnicity). Education takes place in three stages: Primary education, Secondary education, and Post-secondary education. Mathematics is a compulsory subject for students in the primary (grades 1–6) and secondary schools (grades 7–10 or 7–11, depending on the course of study). In all public schools (which are 95% of all MOE registered schools), most teachers have been trained at the sole institute for teacher education in Singapore, the National Institute of Education (NIE). The NIE is an institute of the Nanyang Technological University and it functions like a College of Education of the university. In the rest of the 5% of MOE registered schools, which are mainly independent private schools, like the Singapore American School and the Korean International School, teachers need not be NIE trained.

As NIE is the sole institute for teacher education, it is *Singapore's education think-tank* (Tan et al., 2017, p. 3) and works with its partners, the Ministry of Education (MOE) and schools, to shape the future of education. Unlike many other countries, in Singapore the MOE takes sole responsibility over the recruitment, preparation, certification, appointment, and deployment of teachers, who are public servants, for schools (Low & Tan, 2017). It does this in partnership with NIE, schools, and other stakeholders (e.g., teachers, parents, other government ministries, universities, and the private sector). The MOE recruits teachers from the top one-third of each cohort of the graduating class who qualify for tertiary education; only one of eight applicants interviewed is accepted. Apart from satisfying basic academic standards, aspiring teachers also must have the aptitude and interest in teaching. These are ascertained through interviews conducted by a panel of educators, namely MOE officials, school leaders, and teachers (Teo, 2000). Following recruitment, preservice teachers enrolled in the four-year degree programmes are given a scholarship for their university education leading to a Bachelor's degree with a teaching qualification. They do their university education at the NIE. Upon successful completion of their four years of study, they are employed as full-time civil servants, known as General Education Officers (GEOs). Those preservice teachers already holding university degrees are enrolled in the postgraduate diploma programmes and are immediately employed as GEOs. They receive a monthly salary, including all other benefits enjoyed by civil servants in the country. These GEOs are sent to NIE for their preservice teacher education. Their tuition fee for study at the NIE is paid by the Ministry. This significant capital investment by the government, for preservice teachers in both the four-year degree programmes and postgraduate diploma programmes, calls for a bond of service by the student teachers upon graduation from NIE; the bond ranges from 3 to 4 years depending on the programme of study. In this chapter, we describe the rigorous preservice education of mathematics teachers.

INITIAL MATHEMATICS TEACHER EDUCATION

There are two programmes for preservice mathematics teachers. The preferred programme is the Bachelor of Arts (respectively, Science) with Education, which we normally call BA/BSc (Ed). Student teachers are accepted into the programme based on their A-Level[1] exam or polytechnic results. After Year 10, students in Singapore either opt for a two-year junior college education and take the A-Level, or go to a polytechnic school for a more specialized diploma.[2] Both are viable routes towards a university education. The Singapore-Cambridge General Certificate of Education Advanced Level (GCE A-Level) examination is an annual national examination that is taken by Year 12 students in the junior colleges. Polytechnics in Singapore are three-year diploma-granting tertiary institutions. The polytechnics are not universities. Although the polytechnic diplomas that are conferred can often be used to obtain course and year exemptions in related fields in many universities locally and around the world, no such exemptions are permitted for the BA/BSc (Ed) programme.

The second programme for preservice mathematics teachers is the Postgraduate Diploma in Education. The student teachers in this programme would have obtained relevant degrees from other universities prior to their admission. Upon completion of the degree or 16-month postgraduate diploma programme in NIE, the teacher candidates are automatically certified to teach in Singapore schools. The subject they teach depends on their content specialization in their study programmes.

Singapore recognizes teacher qualifications from other countries on a case-by-case basis. The NIE is the sole teacher accreditation institution in Singapore. However, private schools are not required to employ only NIE accredited teachers. The NIE programmes prepare mathematics teachers for changing school environments through implementing research-based curricula, input from MOE, and regular curriculum and programme reviews. The Ministry of Education works very closely with NIE to prepare teachers for Singapore schools. The school mathematics curriculum is revised once every 6 years by the MOE and NIE adjusts its teacher preparation curriculum accordingly to ensure that teachers are current in their knowledge and practices. In addition, major education shifts are usually implemented close to their principles by schools, education officers, and the NIE.

---

[1] The A-Level exam is jointly administered by the University of Cambridge Local Examinations Syndicate and the Singapore Ministry of Education. It is similar to the A-Levels in the United Kingdom and typically consists of exams in 4 to 6 subjects.

[2] Specialised diplomas range across many specialisations and include business, computer science, engineering, and nursing.

Undergraduate Programme [BA/BSc (Ed)]

Good results in the A Levels or polytechnic exams are required for entry into the undergraduate programme. Candidates also need to pass a stringent interview with the MOE. The BA/BSc (Ed) is a 4-year direct honors programme consisting of about 130 Academic Units (AU). An AU translates to one hour of contact time per week for 13 weeks. Prospective mathematics teachers can choose between the Primary and Secondary tracks. The two tracks are considered non-transferable; i.e., a graduate from the primary track cannot teach in a secondary school and vice-versa. An inservice teacher who wants to change tracks needs to undergo a conversion programme to the appropriate track.

In the secondary track, prospective mathematics teachers take two Academic Subjects (AS), one of which must be mathematics. AS subjects are taught at two levels. The higher level (AS1) consists of 18 courses while AS2 consists of a subset of 5 courses, which can be considered to amount to a minor in mathematics. Thus, those who take AS2 mathematics are often restricted to teaching the lower levels of secondary school mathematics after graduation. AS (AS1 and AS2) Mathematics consists of typical mathematics undergraduate courses, such as Calculus, Linear Algebra, Finite Mathematics, Combinatorics, Real Analysis, Number Theory, Statistics, Galois Theory, and Graph Theory. The five courses taken by AS2 mathematics students are all Year 1 courses: Calculus I, Linear Algebra I, Finite Mathematics, Mathematical Problem Solving, and Number Theory. As an indication of the scope of a course, the topics covered in Finite Mathematics are detailed as follows: basic principles of counting, permutations and combinations, distributions, generalized permutations and combinations, Binomial Theorem and combinatorial identities, the Pigeonhole Principle, sample space and discrete probability distributions, conditional probability, and independent events.

Secondary track student teachers take two Curriculum Studies (CS) series of courses that correspond to their AS programme and level. For example, a prospective mathematics teacher doing AS1 Mathematics and AS2 English Language will take CS1 Mathematics and CS2 English Language. CS1 (major teaching subject) consists of four content-specific pedagogy courses; CS2 (minor teaching subject) has three. CS mathematics courses cover mathematics learning theories, assessment construction, microteaching, and developing pedagogical content knowledge for specific secondary school mathematics topics such as arithmetic, basic algebra, matrices, sets, probability, and statistics. The courses focus on developing a deep understanding of the mathematical concepts in these topics, and an inquiry mindset towards improving teaching. In addition, a variety of issues related to classroom teaching and learning mathematics in Singapore are discussed: the Singapore Mathematics Curriculum Framework, aims of mathematics education, mathematical problem solving, learning theories relevant to mathematics education, and lesson planning. Student teachers will find many opportunities to develop crucial craft skills for teaching mathematics: explaining concepts, demonstrating examples, selecting and sequencing questions, and using appropriate

Information and Communication Technology (ICT) tools. The extra course for CS1 includes topics that are taught in Additional Mathematics for students who take the General Certificate in Education at the Ordinary Level examination at the end of the first four years of their secondary schooling (Year 10). Additional Mathematics is taught over two years (Years 9 and 10) and includes more difficult topics, such as exponential and logarithmic functions, the binomial theorem, and calculus. Students intending to major in Science are strongly encouraged to take Additional Mathematics as a foundation for A-Level Mathematics.

In addition to content and CS courses, the third component of the programme is called Education Studies. This consists of about six general pedagogy courses covering Educational Psychology, Classroom Management, Use of Technology in Teaching, and Learning Theories. The students have 22 weeks of Practicum, split as 2-5-5-10 across their four years of study. Figure 5.1 contains a summary of the courses taken by a hypothetical AS1 mathematics student, who has chemistry as a minor teaching subject, and a hypothetical AS2 mathematics student, who has chemistry as a major teaching subject.

Primary track student teachers take one AS subject and two CS subjects. Thus, those who take Mathematics as their (only) AS subject would take it at AS1 level. These would become mathematics specialists and are even stronger in content mathematics than their counterparts in the Secondary track who take AS2 Mathematics. Unlike their Secondary track counterparts, Primary track student teachers may choose to take CS Mathematics even though they do not take AS Mathematics. Students without CS Mathematics would not teach mathematics when they graduate. There is another suite of courses, called Subject Knowledge in Mathematics, which CS Mathematics students have to take. These courses are designed to help teachers improve their teaching by giving them a perspective of primary mathematics from a higher standpoint, a view espoused by Felix Klein (Klein, 1924a, b). The courses are Number Topics, Geometry Topics, and Further Mathematics Topics (which comprises topics such as algebraic problem solving and some elementary statistics). All CS Mathematics students take the first two. CS Mathematics students without AS Mathematics will also take the third course. The main topics for Number Topics are numeration systems, number operations, divisibility, and ratios, proportions, and rates. The main topics for Geometry Topics are parallels and polygons, congruence, triangles and quadrilaterals, length, area and volume, and symmetry and tessellations.

Postgraduate Diploma in Education

A good undergraduate degree is required for entry into the Postgraduate Diploma in Education (PGDE) programme. Undergraduate degrees are considered good only if they were Honors (cum laude or above) degrees and come from reputable English-language universities. In addition, secondary track applicants need to be mathematics majors or have read a substantial number of mathematics courses in their undergraduate programme. Primary track applicants need not be

Programme Component	AS1 Mathematics Student (with Chemistry as AS2)	AS2 Mathematics Student (with Chemistry as AS1)
AS1	Calculus I	Calculus I
	Linear Algebra I	Linear Algebra I
	Finite Maths	Finite Maths
	Number Theory	Number Theory
	Mathematical Problem Solving	Mathematical Problem Solving
	Computational Mathematics	
	Calculus II	
	Linear Algebra II	
	Statistics I	
	Differential Equations	
	Complex Analysis	
	Statistics II	
	Real Analysis	
	Modern Algebra	
	Combinatorial Analysis	
	Metric Spaces	
	Galois Theory	
	Academic Exercise*	
AS2	4 Chemistry courses	17 Chemistry courses
CS1	Teaching and Learning Maths 1	Teaching and Learning Maths 1
	Teaching and Learning Maths 2	Teaching and Learning Maths 2
	Teaching and Learning Maths 3	Teaching and Learning Maths 3
	Specialized Areas in Teaching and Learning Mathematics	
CS2	3 Teaching Chemistry courses	4 Teaching Chemistry courses
Educational Studies	Educational Psychology I	Educational Psychology I
	Educational Psychology II	Educational Psychology II
	Teaching and Managing Learners at the Primary Level (or Teaching and Managing Learners at the Secondary Level)	Teaching and Managing Learners at the Primary Level (or Teaching and Managing Learners at the Secondary Level)
	Technologies for Meaningful Learning	Technologies for Meaningful Learning
	Assessing Learning and Performance	Assessing Learning and Performance
	The Social Context of Teaching and Learning	The Social Context of Teaching and Learning
	Character and Citizenship Education	Character and Citizenship Education
	Multicultural Studies: Appreciating & Valuing Differences	Multicultural Studies: Appreciating & Valuing Differences
Practicum	Practicum (22 weeks)	Practicum (22 weeks)

FIGURE 5.1. Summary of Courses for Students in Two Levels of Mathematics Programmes.
*The Academic Exercise is a 6 AU course that acts as a capstone for the AS studies. It involves research, writing of a report/dissertation, and seminar presentation on a topic in mathematics.

mathematics majors. Candidates also need to pass a stringent interview with the MOE, and typically spend about half a year doing contract teaching in a Government school to ascertain for themselves if teaching as a career would fit them. They work as supervised untrained teachers and are actual teachers of record at the school. If they perform satisfactorily, they are enrolled as full-time students at NIE.

The PGDE is a 16-month programme consisting of about 40 AUs. As in the undergraduate programme, prospective mathematics teachers can choose between the Primary and Secondary tracks. Unlike the NIE undergraduate programme, there is no AS component in the PGDE. Student teachers in the Secondary track must have the required subject expertise from their previous undergraduate studies, whereas Primary track teachers need not have subject expertise at undergraduate level. In particular, Primary track mathematics teachers are usually not mathematics majors. All student teachers take two CS subjects. Both CS subjects consist of about four pedagogy courses, similar to those in the BA/BSc (Ed) programme. The Educational Studies component of the PGDE programme is also similar to that of the undergraduate programme. The students go for 14 weeks of Practicum, split as 4-0-10 across three semesters of study. Unlike the undergraduate programme, non-maths majors in the Primary track take only two Subject Knowledge courses, Number Topics and Geometry Topics, because of a lack of curriculum time.

Theory-Practice Nexus

In the Practicum, student teachers are assigned a School Coordinating Mentor (SCM), a Cooperating Teacher (CT), and a NIE Supervisor (NIES). The role of the SCM is bi-fold: an administrative one, taking care of all practicum matters at the school level and also liaising with the NIES on matters that warrant the NIES's attention; and a professional one, overseeing the mentoring of the student teacher. The CT is a mentor who works closely with the student teacher during Practicum. Student teachers are attached full-time to a school for the period of the Practicum. Schools select their SCMs and CTs judiciously to give and bring out the best in student teachers during Practicum.

In the undergraduate programme, the first and second attachments occur at the end of the first and second years of study, respectively. These are 2-week and 5-week Teaching Attachments. The student teacher observes lessons and helps plan lessons. They write their reflections and share them with the CT, SCM, and the NIES. In the second year, they may co-teach a class with the CT. The NIES observes the class and gives feedback. The last two attachments occur at the end of the third and fourth years of study, respectively. These are 5 and 10-week Teaching Practicums. Student teachers are assigned to classes, which they teach on their own. The CT observes about 6 lessons and the SCM may co-observe a few lessons with the CT. NIES observes 2 lessons. The CT, SCM, and NIES each give feedback to the student teacher and the observation reports form part of the overall assessment of the student teacher.

In the PGDE programme, the first attachment occurs at the end of the first semester of study. This is a 4-week Teaching Practicum. Student teachers are assigned to classes, which they teach on their own. They write their reflections and share them with the CT and the NIES. Once a week, they return to join their CS course in NIE, where they share their experiences and discuss how theory works in actual practice. The final attachment occurs at the end of the programme. This is a Teaching Practicum of 10 weeks. Student teachers are assigned to classes, which they teach on their own. The CT observes about 6 lessons, the SCM may co-observe a few lessons, and the NIES observes only 2 lessons. They each give feedback to the student teacher, and the observation reports form part of the overall assessment of the student teacher.

The Mathematician Educator

Grossman et al. (1989) stated that it had become "increasingly clear that … [teacher educators] can no longer assume that the subject matter component of teacher preparation is fulfilled by undergraduate coursework in other departments" (Grossman et al., 1989, p. 24). Mathematics teacher education in NIE is quite different from many other teacher education universities where the mathematics education department and the mathematics department are distinctly different. At NIE, active research mathematicians work together with mathematics educators to prepare school mathematics teachers (Tay et al., 2019). The work of both groups of faculty are complementary. The common direction is summed up in the Vision and Mission statements of the department as follows:

Vision: To be a world-class symbiosis of mathematicians and mathematics educators.
Mission: Through research and exemplary teaching, nurture mathematician educators for Singapore and the world.

A *mathematician educator* is a teacher of mathematics who bases his/her pedagogy on a strong foundation of mathematics disciplinarity, learning theories, and mathematics education research findings. To bridge a common divide between mathematicians, who seem to teach the *subject*, and mathematics educators, who seem to teach the *child*, a mathematician educator is envisaged as an amalgam of both. His/her knowledge base is of both mathematics and its disciplinarity, and the student and his/her psychology. There are no academic qualifications to certify one as a mathematician educator—one can be a university professor or a kindergarten teacher. When university professors adjust their teaching for different students and use suitable examples with an understanding of pedagogical content knowledge, then they are mathematician *educators*. When schoolteachers understand they can prove that three tangram sets cannot be assembled into a square (instead of knowing only that students, when challenged, will not be able to do so), then they are *mathematician* educators. We propose the term *mathematician educator* to emphasise the need for both aspects to be embodied in the one person. The term *mathematics educator*, when wrongly emphasised, would imbue the human personality only on the *educator* aspect.

The BA/BSc (Ed) programme is the reason for the amalgam of mathematicians and mathematics educators. Research mathematicians are hired to teach the AS component of the degree programme. The Mathematics and Mathematics Education department (MME) consists of 13 faculty with doctorates in mathematics and 12 with doctorates in mathematics education. Most of the mathematicians are active in mathematics research, with respectable Google Scholar citations.[3] For example, one of the Associate Professors has a citation index of 841 with *h*-index 13. He is also regarded as one of the top researchers in the area of chromatic polynomials. Some of the mathematicians work enthusiastically in both mathematics and mathematics education research. Another Associate Professor of this hybrid tendency has a Google citation index of 638 with *h*-index 17. The mathematics education faculty are equally productive in research. A Professor in Mathematics Education in the department has a Google citation index of 2147 with *h*-index 23. The mathematics educators teach the CS components of both the degree and PGDE programmes.

In practice, both groups of faculty often interact in seminars and in informal gatherings. Discussions among them range from teaching issues (and thus, mathematics educators give input to their colleagues regarding pedagogical issues in

---

Vignette 5.1. A Mathematician Shares: Impossible is Nothing but "Nothing" May be Possible!

In the teaching of probability at the secondary school level, it is common to see teachers say that an event is impossible if and only if its probability is 0. While the "only if" direction is true, a mathematics educator was shocked to learn from his colleague that the other direction is not. The mathematician explained that in Measure Theory, it is possible to give a measure of 0 to an element in a set, ergo the element exists and is not impossible, but it has measure 0. Probability is a measure of chance. For example, in an experiment where a person randomly chooses a number on the number line between 0 and 1, the probability of choosing any number x is 0. That does not mean that choosing x is impossible. Finally, the mathematician convinced his colleague by guiding him to agree that a point has no length but it exists, a line has no area but it exists, and a plane has no volume but it exists. The mathematics educator would now be able to teach the correct concept in the CS courses to the teachers.

---

[3] Individual Google Scholar citations are accessible for researchers who make their profiles public. These can be seen by searching out names on https://scholar.google.com/.

Vignette 5.2. Mathematics Educators Share Effective Pedagogical Techniques: A Mathematician Won a Teaching Commendation!

A mathematician who was recruited from another Singapore university found that his teaching feedback scores were rather low. He had been awarded teaching commendations in his previous university but discovered that, compared to the excellent teaching of his colleagues in NIE, he was ranked low. The mathematician had an open mind and discussed teaching strategies with other mathematician and mathematics education colleagues. He adopted and adapted cutting edge pedagogical methods, and over time, his feedback scores improved until he also finally won a teaching commendation.

---

undergraduate mathematics) to mathematical problems and concepts (and here, mathematicians explain problem solving techniques and conceptual underpinnings of secondary mathematics, such as that of limits in calculus and measure in probability). Sometimes in the Master of Education programme, mathematicians team with mathematics educators to co-teach courses such as Number Theory and the Teaching of Arithmetic. The symbiosis ensures that the curriculum is mathematically sound and the pedagogy realistic and research-based. Vignettes 5.1 and 5.2 show the symbiotic relationship between the two groups of colleagues.

SYSTEMATIC ON-GOING REVIEW AND REVISION OF CONTENT AND PEDAGOGY COURSES FOR TEACHER EDUCATION

NIE carries out regular reviews of its curriculum at different levels. These lead to revisions and development of various programmes to respond to the changing education landscape. Nonetheless, in all these reviews, the history of current practices is taken into consideration; this accounts for the evolutionary character of the changes resulting from the reviews. For example, at the institute level, a major revision resulted in the *Singapore Teacher Education for the 21ˢᵗ Century* (TE²¹) report in 2009 (NIE, 2009). In this report, an existing framework of Values, Skills, and Knowledge (VSK) was given an important touch-up in view of the changing moral landscape. For example, some teachers found it difficult to separate their professional and private selves in Facebook postings and allowed their students access to less modestly attired pictures of themselves. The new framework, V³SK, made special focus on values as an important component of teacher education. Here, the component of Values was further explained as *Learner-Centred Values, Teacher Identity*, and *Service to the Profession and Community* (NIE, 2009, p. 45). Learner-centered values put the student, as compared to the subject, as the centre of teaching and learning.

Examples of learner-centered values are empathy, a belief that all students can learn, a commitment to nurturing the potential of each child, and valuing diversity.

Experienced educators are aware that some teachers burn out because they give everything of themselves to their students, ignoring their own needs and growth. The second facet of Values reminds the teacher of his/her own self as a person and as a teacher professional. Examples of teacher identity values are aiming for high standards, having an enquiring nature, having a quest for learning, striving to improve, being adaptive and resilient, and being ethical and professional.

Finally, teachers must know that their profession, though most impactful, is not confined to the classroom. They must be aware that they are part of a community, of other educators, and of society. Examples of these values are collaboration in learning and practice, building apprenticeship and mentorship, having social responsibility and engagement, and being a steward of society's values. Within this framework, student teachers would view their learning and development from the tripartite perspectives of the student, oneself as a teacher, and one's colleagues and community. The review recommended that aspects of all three major components should be considered in downstream reviews of curricula and courses.

In-House Review of Courses After Every Cycle of Implementation

At the department level, a comprehensive curriculum review and development for the AS courses of the undergraduate programme has been ongoing since 2015. All mathematics faculty, together with some maths education faculty, were involved in the review following Tyler's model for curriculum development (Tyler, 1949). Tyler's model is a basic structure for developing and evaluating instruction consisting of four aspects: defining learning objectives; designing learning experiences; organizing learning experiences; and evaluating learning and revising the curriculum. The experience of many university mathematics teachers is that, upon joining the faculty, they would be assigned a course to teach, and they would teach it in their own ways based on the course description. Many would not know what other faculty were doing in their classes. The undergraduate curriculum would thus comprise an eclectic mix of content and teaching styles. Tyler's model was introduced to MME faculty and a holistic approach based on Tyler's was agreed upon as the basis for reviewing and developing the undergraduate mathematics curriculum.

A summary of the learning objectives of the BA/BSc (Ed) mathematics programme is as follows.

At the end of the undergraduate mathematics programme, the learner should:
- Possess a solid foundation of school mathematics and the canons of undergraduate mathematics (Content);
- Possess the cognitive ability to understand mathematical ideas, make mathematical connections, and reason mathematically (Cognition);
- Possess the ability to pose, solve, and extend mathematical problems (Problem Solving);

- Possess computational fluency that supports mathematical thinking (Computation);
- Possess communication skills that enable them to convey sound mathematics confidently (Communication); and
- Possess a productive disposition towards mathematics (Disposition).

Towards these ends, courses were reviewed, and suitable learning experiences and assessments were recommended so that a holistic approach across all courses in the programmes would be adopted. For example, a sub-goal of the Cognition component is "be able to read mathematical text or language with understanding." Courses beginning with Linear Algebra 1 in Year 1 to the capstone Academic Exercise would slowly build up the learner's ability in this area with learning experiences and activities, and assessments of increasing difficulty. This development ranges from *reading a definition* in Linear Algebra I to *reading a proof* in Calculus II in Year 2 to *reading a mathematics paper and writing a report* in the Academic Exercise, as shown in Vignette 5.3. The Academic Exercise begins at the end of Year 3 and stretches to the beginning of Semester 2 Year 4, a period of about seven months. It provides students an opportunity to engage in independent learning and research under the guidance of academic staff. A written report of 50–80 pages and a 40-minute oral presentation are assessed at the end of the study.

Constant Updates to Keep Abreast of MOE Initiatives and School Mathematics Curriculum Revisions

The CS modules, which include knowledge of curriculum, are constantly updated to keep abreast of MOE initiatives and school mathematics curriculum revisions. This ensures that beginning teachers are current in their knowledge of curriculum as well as the overall direction of the education system. For example, in 1997, when the Thinking Schools, Learning Nation initiative was implemented by the MOE, a revision of the school mathematics curriculum was carried out to

---

Vignette 5.3.   Enhancing Rigor: Slowly but Steadily

In more detail, examples and non-examples to demonstrate a new definition were initially given by the lecturer in Linear Algebra I. To make it a personal learning experience for the student, the lecturer would, over a few lessons, gradually switch from giving the examples and non-examples himself to asking students to work out examples and non-examples on their own and express them in class. Instead of explaining all new definitions in class, the lecturer would introduce definitions that were part of the syllabus by setting them as tutorial questions, thus necessitating the students to read the definitions on their own. The exam paper at the end of the course would include at least one new definition in at least one problem to be solved.

ensure that teachers placed emphasis on thinking skills as heuristics for learning mathematics and also problem solving (Kaur, 2019).

Impact of Research on Mathematics Teacher Preparation

Research findings from international studies, such as the Teacher Education and Development Study in Mathematics (TEDS-M; Kaur, Zhu, & Cheng, 2019; Tatto et al., 2008), are carefully examined for international benchmarks. Singapore participated in the TEDS-M study. Preservice teachers preparing to teach both primary school mathematics and secondary school mathematics who participated in the study were ranked amongst the top four participating countries for their Mathematics Content Knowledge and Mathematics Pedagogical Content Knowledge. These findings affirmed that the stringent entry requirement of mathematics content knowledge for prospective mathematics teachers was valid and must be upheld. In addition, the robust development of mathematical pedagogical content knowledge was affirmed. A mismatched finding from the questionnaires administered to the preservice teachers and their mathematics educators was noted. The educators felt that they had provided fairly frequent opportunities for the student teachers to engage in interactive learning experiences, such as to ask questions, participate in class discussion, work in groups, and make presentations to the class. However, with the exception of group work, the preservice teachers rated the other three interactive experiences lower than their educators. This finding provided the educators an issue they had to address in their work with preservice teachers.

National level studies, such as the study of school mathematics curriculum enacted by competent and experienced teachers in Singapore secondary schools (Kaur et al., 2018), also provide input into both preservice and inservice education programmes for mathematics teachers at the NIE. Findings from the study, framed as "Twelve Questions on Mathematics Teaching: Snapshots from a Study of the Enacted School Mathematics Curriculum in Singapore" (Kaur, Toh, et al., 2019), have provided knowledge about pedagogies adopted by competent and experienced teachers worthy of widespread emulation in Singapore schools. The questions range from math talk to use of instructional materials in making mathematical connections. Figure 5.2 shows the content page of "Twelve Questions on Mathematics Teaching." The book is available at www.tinyurl.com/enact-12q.

Educators teaching the courses also carry out independent research to assess the impact of the courses on preservice mathematics teachers (see Kaur, 2017). All of the above research inputs help in advancing evidence-based practice from and for the education of mathematics teachers in Singapore. Figure 5.3 shows how research findings have provided added perspectives for an inservice programme course and a Master of Education (Mathematics) programme course.

CONTENTS

$a^2 + b^2 = c^2$

FIGURE 5.2. Content page of the book entitled *Twelve Questions on Mathematics Teaching: Snapshots from a Study of the Enacted School Mathematics Curriculum in Singapore* (Reprinted with permission)

Management and Leadership Studies. Course 117—Secondary Mathematics

This course comprises six seminars, each 3 hours in duration. The themes of the six seminars are:

1. Framework of school mathematics curriculum;
2. Mathematical problem solving;
3. Classroom pedagogy;
4. The intended and implemented curriculum;
5. Excellence in the mathematics classroom—teacher is key; and
6. Assessment—Teaching and Learning (Traditional and Alternative Modes).

Though the focus of the seminars has remained the same over the past decade, the readings that participants undertake in preparation for the seminars have evolved and have increasingly included current research carried out by mathematics educators in Singapore. Two key sources of readings for the participants of the July 2019 and January 2020 intakes have been the books:

- Toh, T. L., Kaur, B., & Tay, E.G. (2019). *Mathematics education in Singapore*. Singapore: Springer.
- Kaur, B., et al. (2019). *Twelve questions on mathematics teaching: Snapshots from a Study of the Enacted School Mathematics Curriculum in Singapore*. Singapore: National Institute of Education.

Master of Education (Mathematics). Course 901—Theoretical Perspectives and Issues in Mathematics Education Research

This course comprises 13, three-hour seminars and an additional 13 hours of e-learning. The readings for the course are updated periodically, for every cycle of implementation, to reflect currency in the field and research in the Singapore context. At present almost 40% of the readings for the course are on research carried out on mathematics education in Singapore by educators from the NIE. Due to the limitation of space, only five of the readings are listed here.

- Toh, T. L., Kaur, B., & Tay, E. G. (2019). *Mathematics education in Singapore*. Singapore: Springer. [There are 21 chapters in this book based on research in Singapore.]
- Lee, N. H., Yeo, D. J. S., & Hong, S. E. (2014). A metacognitive-based instruction for primary four students to approach non-routine mathematical word problems. *ZDM Mathematics Education, 46*, 465–480.
- Kaur, B. (2011). Enhancing the pedagogy of mathematics teachers (EPMT) project: a hybrid model of professional development. *ZDM Mathematics Education, 43*, 791–803.
- Kaur, B. (2019). The why, what and how of the 'Model' method: A tool for representing and visualizing relationships when solving whole number arithmetic. *ZDM Mathematics Education, 51*, 151–168.
- Kaur, B. (2017). Impact of the course teaching and learning of mathematics on preservice grades 7 and 8 mathematics teachers in Singapore. *ZDM Mathematics Education, 49*, 265–278.

FIGURE 5.3. Infusion of Local Research Into Education of Mathematics Teachers

CERTIFICATION AND SYSTEMIC
SUPPORT FOR BEGINNING TEACHERS

Upon graduation from the NIE, all teachers, including mathematics teachers, are certified to teach in public schools in Singapore. There are no additional testing requirements. However, they are on probation during their first year of teaching and need to pass a confirmation by the end of the year. The confirmation exercise is school-based and involves school leaders observing the teacher's work in school and grading it as satisfactory or otherwise. In rare cases, if a teacher fails to pass it, the school may decide to give him/her another semester to pass. Should a teacher fail to be confirmed, he/she has to pay the MOE a liquidated damage (compensation) that is a percentage of the cost borne by the MOE for their teacher education, including all monies paid to them either as salary or stipend (if they were on a four-year scholarship).

NIE graduates are posted by the MOE to schools to teach their respective subjects. Normally, distance from home to school and the needs of schools are two basic criteria that decide where a teacher is posted. All teachers, irrespective of their seniority or specializations, have the same working conditions, such as school hours, school vacation, and professional development opportunities necessary for their on-going development under the guidance of their mentors. Beginning teachers are inducted into the profession at the end of their preservice education at the NIE. This is done by the MOE and the Academy of Singapore Teachers (AST) through briefing sessions related to the ethos of the teaching profession. The five facets of the Ethos of the Teaching Profession are:

- *Our Singapore Educators' Philosophy of Education*, which captures the core beliefs and tenets of the teaching profession and serves as the foundation of teachers' professional practice;
- *The Desired Outcomes of Education*, which establishes a common purpose for the teaching fraternity, guiding educational and school policies, programmes and practices;
- *The Teachers' Vision*, which articulates the aspirations and roles of the teaching profession, helping teachers to focus on what to do in pursuit of professional excellence;
- *The Teachers' Pledge*, which constitutes an act of public undertaking that each teacher takes to uphold the highest standards in professional practice; and
- *The Teachers' Creed*, which codifies the practices of retired and present educators and makes explicit their tacit beliefs. It provides a guide for teachers to fulfil our responsibilities and obligations, and to honor the promise of attaining professional excellence. (https://academyofsingaporeteachers. moe.edu.sg/professional-excellence/ethos-of-the-teaching-profession)

At the school level, beginning teachers are assigned less than the full load of teaching in their first year, normally 80 percent of instruction time compared to the other teachers senior to them in the school. They work with their mentor, who is a senior teacher, to plan and enact their lessons. In addition, they are members of Professional Learning Communities (PLC) involving mathematics teachers in the school and are encouraged to attend professional development deemed necessary by their mentors. In their school PLC, they work alongside mathematics teachers examining issues the community has agreed to work on for the year. They are gradually given curricular and co-curricular tasks to work on under supervision. By the second year of their teaching, they have a full load of teaching and specific co-curricular assignments, such as overseeing and/or planning activities in the school or in the community that students engage with after their curriculum hours.

CHALLENGES AND IMPETUS TO REMAIN NATIONALLY ROOTED WHILST GLOBALLY RELEVANT

The MOE recruits teachers on a need basis, sends them to NIE for preservice education, and subsequently deploys them to schools. This process ensures, within reasonable margins, surpluses and shortages of mathematics teachers in Singapore schools. However, some challenges will always prevail.

In the past, some of the challenges faced in recruitment of good prospective mathematics teachers stemmed *from teaching being not a choice profession* because the remuneration for beginning teachers was not at par with beginning engineers, accountants, and doctors. But at present with the remuneration beginning teachers receive, it is possible to attract prospective teachers from the top one-third of cohorts of graduates. In essence, good teachers command good remuneration that is at par with other professions, such as engineers, and administrative and executive officers in the public sector.

In the preparation of mathematics teachers, there needs to be a continued focus on mathematics content knowledge of teachers alongside pedagogical content knowledge. The present faculty of the Mathematics and Mathematics Education Academic group at NIE is best suited to ensure this as both mathematicians and mathematics educators with sound background in mathematics are working alongside each other. However, should there be any attempt to prepare mathematics teachers otherwise, we envisage challenges in ensuring the mathematical competency of teachers in our system. For example, NIE is part of Nanyang Technological University (NTU), which has another mathematics department in its College of Science. For logistical efficiency, some quarters have raised the possibility of transferring all teaching of AS courses to faculty in the other NTU mathematics department.

At the school level, mentors to beginning mathematics teachers also need to be competent teachers themselves. However, at times due to the absence of such teachers, our beginning mathematics teachers do not get the best possible support. However, a solution is appearing on the horizon. Presently, as of January 2020, the mathematics chapter at the AST has begun facilitating meetings amongst mathematics senior teachers, who would offer mentorship to beginning teachers in their cluster of schools. Unlike other diverse multi-ethnic communities, the English language of instruction does not pose a serious problem in Singapore mathematics classrooms as all classroom instruction is in English. Beginning teachers, however, often have difficulties with classroom management, differentiated teaching, and assessment. The AST and NIE, through their various professional development programmes, do try to level up the knowledge of every teacher, but levelling up school-based hands-on experience remains a challenge.

REFERENCES

Grossman, P. L., Wilson, S. M., & Shulman, L. S. (1989). Teachers of substance: Subject matter knowledge for teaching. In M. C. Reynolds (Ed.), *Knowledge base for the beginning teacher* (pp. 23–36). Pergamon Press.

Kaur, B. (2017). Impact of the course teaching and learning of mathematics on preservice grades 7 and 8 mathematics teachers in Singapore. *ZDM Mathematics Education, 49*, 265–278. DOI 10.1007/s11858-016-0830-8

Kaur, B. (2019). Overview of Singapore's education system and milestones in the development of the system and school mathematics curriculum. In T. L. Toh, B. Kaur, & E. G. Tay (Eds.), *Mathematics education in Singapore* (pp. 13–33). Springer Nature. DOI 10.1007/978-981-13-3573-0_2

Kaur, B., Tay, E. G., Toh, T. L., Leong, Y. H., & Lee, N. H. (2018). A study of school mathematics curriculum enacted by competent teachers in Singapore secondary schools. *Mathematics Education Research Journal, 30*, 103–116. DOI 10.1007/s13394-017-0205-7

Kaur, B., Toh, T. L., Lee, N. H., Leong, Y. H., Cheng, L. P., Ng, K. E. D., & Yeo, K. K. J., Yeo. B. W. J., Wong, L. F., Tong, C. L., Toh, W. Y. K., & Safii, L. (2019). *Twelve questions on mathematics teaching: Snapshots from a Study of the Enacted School Mathematics Curriculum in Singapore*. National Institute of Education.

Kaur, B., Zhu, Y., & Cheang, W. K. (2019). Singapore's participation in international benchmark studies—TIMSS, PISA and TEDS-M. In T. L. Toh, B. Kaur, & E. G. Tay (Eds.), *Mathematics education in Singapore* (pp. 101–137). Springer Nature. DOI 10.1007/978-981-13-3573-0_6

Klein, F. (1924a). *Elementary mathematics from an advanced standpoint: Arithmetic, algebra, analysis*. Dover Publications.

Klein, F. (1924b). *Elementary mathematics from an advanced standpoint: Geometry*. Dover Publications.

Low, E-L., & Tan, O-S. (2017). Teacher education policy: Recruitment, preparation and progression. In O-S. Tan, W-C. Liu, & E-L. Low (Eds.), *Teacher education in the 21st century—Singapore's evolution and innovation* (pp. 11–32). Springer Nature. DOI 10.1007/978-981-10-3386-5_2

National Institute of Education (NIE), Singapore. (2009). *A teacher education model for the 21st century*. National Institute of Education.

Tan, O-S., Liu, W-C., & Low, E-L. (2017). Teacher education futures: Innovating policy, curriculum and practices. In O-S. Tan, W-C. Liu, & E-L. Low (Eds.), *Teacher edu-*

cation in the 21st century—Singapore's evolution and innovation (pp. 1–9). Springer. DOI 10.1007/978-981-10-3386-5_1

Tatto, M. T., Schwille, J., Senk, S. L., Ingvarson, L., Peck, R., & Rowley, G. (2008). *Teacher Education and Development Study in Mathematics (TEDS-M): Policy, practice, and readiness to teach primary and secondary mathematics. Conceptual framework.* International Association for the Evaluation of Educational Achievement (IEA).

Tay, E-G., Ho, W-K., Cheng, L-P., & Shutler, P-M-E. (2019). The National Institute of Education and mathematics teacher education: Evolution of preservice and graduate teacher education. In T-L. Toh, B. Kaur, & E-G. Tay (Eds.), *Mathematics education in Singapore* (pp. 351–391). Springer. DOI 10.1007/978-981-13-3573-0_15

Teo, C. H. (2000). *Speech by Radm (NS) Teo Chee Hean, Minister for Education and Second Minister for Defence, at the 2nd Teaching Scholarship Presentation Ceremony.* https://www.nas.gov.sg/archivesonline/speeches/record-details/75920429-115d-11e3-83d5-0050568939ad

Tyler, R. W. (1949). *Basic principles of curriculum and instruction.* The University of Chicago Press.

CHAPTER 6

MATHEMATICS TEACHER EDUCATION IN AOTEAROA NEW ZEALAND

Glenda Anthony and Raewyn Eden
Massey University

Tirohia kia mārama
Whawhangia kia rangona te hā

Observe to gain enlightenment
Participate to feel the essence

Initial teacher education in Aotearoa New Zealand is situated in an educational space focused on promoting equity and excellence for all children and young people. Within this space there is an urgent need to support our beginning teachers to be able to give active expression to Te Tiriti o Waitangi and to respond to educational needs in the future. Outlining the nature of mathematics education in the initial education context, this chapter provides examples of how research-based reforms associated with practice-based pedagogies support teacher candidates to learn the work of ambitious pedagogies for the mathematics classroom. Foundational to current revisions in initial teacher education and associated mathematics education courses is the drive towards building teacher capacity in inclusive practices that support positive cultural identity and well-being for all students.

International Perspectives on Mathematics Teacher Education, pages 113–141.

INTRODUCTION

Initial teacher education (ITE) in Aotearoa New Zealand,[1] as in most countries, is an ever-changing space (Tomorrow's Schools Independent Taskforce [TSIT], 2018). It is a space that is dominated in recent times by a multitude of socio-political (Milne, 2016) and educational concerns that present challenges to notions of teacher and teaching quality, student well-being (Berryman et al., 2018), and achievement equity (Cochran-Smith et al., 2016; Hunter & Hunter, 2017). Within ITE, calls for reforms and research priorities have broadened from long-standing concerns related to the practice-theory divide (Whatman & MacDonald, 2017) towards seeking accountability and understandings about "ITE and its relationships to teacher and student learning" (Ell et al., 2017, p. 345). As a consequence, research exchanges and reforms increasingly focus on equity-centered teacher education (Cochran-Smith et al., 2016), the orientation of the teacher towards using culturally responsive actions (Averill & McRae, 2019), teaching as inquiry and development of adaptive expertise (Anthony, 2018), and professional standards for teaching (Sinnema et al., 2017).

In the context of mathematics teacher education, this chapter highlights researched reforms within Aotearoa New Zealand, including research on teacher content and pedagogical knowledge, cultural competencies, practice-based curriculum innovations, and Māori-medium initiatives. To foreground this work, we begin the chapter with a brief overview of education systems in Aotearoa New Zealand and the role of mathematics curricula. We then overview ITE provisions across Aotearoa New Zealand, including the place of mathematics teacher education and a case study of reform. We conclude the chapter with a return to the challenges and visions for mathematics teacher education going forward.

SETTING THE SCENE: EDUCATION IN AOTEAROA NEW ZEALAND

The country's founding document, *Te Tiriti o Waitangi—The Treaty of Waitangi*, positions Aotearoa New Zealand as a bicultural nation with indigenous Māori and non-Māori as equal partners (Tomlins-Jahnke & Warren, 2011). Regarded as "a foundation, both in moral and practical terms, of our schooling system" (TSIT, 2018, p. 33), the three broad principles of partnership, protection, and participation are expected to underpin all school decision making:

> The Treaty of Waitangi principle puts students at the centre of teaching and learning, asserting that they should experience a curriculum that engages and challenges

---

[1] Aotearoa New Zealand is a bilingual title that includes English and te reo Māori, both official languages with equal status. New Zealand is the country's official legal name; however, "Aotearoa" and "New Zealand" are used either interchangeably or together in a wide range of official documents (Hudson, 2019).

them, is forward-looking and inclusive, and affirms New Zealand's unique identity. (Ministry of Education [MoE], 2019d)

Teachers are expected to demonstrate commitment to Te Tiriti o Waitangi. Regardless of their role or teaching context, all teachers are required to have "an understanding of education within the bicultural, multicultural, social, political, economic and historical contexts of Aotearoa New Zealand" and "a knowledge of tikanga and te reo Māori" (*Māori customs and language*) (Education Council, 2017, p. 28). The pursuit of a strong social justice agenda that acknowledges the rights of children and young people within Aotearoa New Zealand is also affirmed in the signatory to three major international human rights statements: the *Convention on the Rights of the Child* (United Nations, 1989); the *Convention on the Rights of Persons with Disabilities* (United Nations, 2007a); and the *Declaration on the Rights of Indigenous Peoples* (United Nations, 2007b).

Aotearoa New Zealand's population is described as super-diverse (Simon-Kumar, 2020). The 2018 census indicated a growth in the overseas-born population; the number of New Zealand residents born outside of the country increased to 27.4%, up from 25.2% in 2013. The growth in the overseas-born population coincides with higher migration over the last five years, especially by young adults coming to study or work in Aotearoa New Zealand. Of note is that Aotearoa New Zealand is home to the largest group of Pacific peoples in the Western world. A multi-ethnic, heterogeneous group, Pacific people include those born in Aotearoa New Zealand and overseas who identify themselves with cultures and languages of Pacific island nations. In 2017, this diversity (students may identify with more than one identity) was represented in our student population: 71% identified as European or Other ethnicity (including New Zealander), 25% as indigenous Māori, 13% as Pacific students, 13% as Asian, and 2% as Middle Eastern/Latin American/African.

For this diverse student group, experiences of poverty (20% living in households of significant hardship), combined with past and current inequities across all levels of education, create significant challenges. "Too many of our Māori and Pacific students, and too many of our students from disadvantaged backgrounds, are not succeeding as they should, are not reaching their potential, and have not been doing so for far too long" (TSIT, 2018, p. 21). In mathematics education, Pacific and Māori students are seriously over-represented in low achievement scores and well-being measures (see Caygill et al., 2016; Chamberlin & Caygill, 2012). Although many seek solutions within the classroom (Averill, 2012; Hunter & Hunter, 2017; Te Maro, 2018; Tweed, 2015), concerns about the capabilities of new teachers to effectively meet the learning needs of students from diverse backgrounds, or those with additional learning needs, are to the forefront of recommended educational reforms (see TSIT, 2018).

The Schooling Sector

Prior to formal schooling, nearly all children attend some form of early child-hood education programme, which is part government-funded. There are many options for such programmes, such as teacher-led, parent-led, and whānau (*family*)-led centres, including over 450 Kōhanga Reo (*Māori-medium centres*). Schooling is compulsory from age 6 to age 16, however nearly all children start at age 5. Primary school (Years 1 to 8) is followed by secondary school (Years 9 to 13). In larger urban centres, options may include an intermediate school for Years 7 to 8. Most students remain at school until they are age 17 (Year 12). Due to the rural and sometimes remote location of some students, the option of distance study is available through Te Aho o Te Kura Pounamu (formerly The Correspon-dence School).

Attending an educational setting that accesses the curriculum through the me-dium of Māori is another important option for students, currently around 2.4% of the school population. For some, this involves bilingual classes within English-medium schools; for others, this involves education in Māori-medium settings, discussed later in more detail.

Our national curriculum comprises The New Zealand Curriculum (MoE, 2007), its parallel document Te Marautanga o Aotearoa, a curriculum based on a Māori worldview (MoE, 2008a, revised 2017a), and Te Whāriki: Early childhood curriculum (MoE, 2017b). The New Zealand Curriculum covers all learning areas and includes values, principles, and key competencies (thinking; using language, symbols, and text; managing self; relating to others; and participating and con-tributing). Together these documents are designed to support schools and early childhood education centres to enact the partnership that is embedded in the Te Tiriti o Waitangi.

Māori Medium Education

Since the 1970s, concerted advocacy within Māori communities has led to a renaissance of Māori language and culture. In the education space, such activism has resulted in the establishment of Māori immersion educational initiatives: Te Kōhanga Reo (*Māori-medium centres for pre-school children*); Kura Kaupapa Māori (*Māori–medium primary schools*); Wharekura (*Māori-medium second-ary schools*); and Whare Wānanga (*indigenous education providers at tertiary level*). In 2017, nearly 20,000 students were enrolled in Māori-medium education, spread across 277 schools and Kura Kaupapa Māori, including English schools with bilingual units (MoE, 2018a). Schools that offer Māori-medium programmes may use Te Marautanga o Aotearoa and use te reo Māori as their language of instruction. Te Marautanga o Aotearoa is focused on skills, competencies, and knowledges connected with the Māori world and necessary to participate in and contribute to Māori society and the wider world (MoE, 2017a).

Developed in response to the aspirations of whānau (*family*) and communities, Māori-medium education is inextricably linked to the regeneration of Māori
culture and language. However, attempts to revitalize te reo Māori continue to be
hampered by shortcomings in teacher supply, curriculum and resource development, and funding of Māori-medium ITE (Trinick, 2019). A further challenge
for mathematics education, in particular, is the need to develop "new linguistic
resources for discussion and dissemination of conceptual materials at high levels
of abstraction" (Stewart et al., 2018, p. 159). Moreover, for many Māori research-
ers/educators, the education provision within kura (*school*) communities remains
a contested space. As noted by Te Maro (2018), kura "must comply to Western
educational institutional structures and bio/knowledge-power systems that are
bounded by politics and culture of colonization that cannot seem to let go of
power" (p. 233).

INITIAL TEACHER EDUCATION: BECOMING A TEACHER

Within a competitive market model of ITE provision, there are currently nine
providers of secondary ITE, 16 providers of primary ITE, and 19 providers of
Early Years ITE programmes. Across the range of providers and programmes,
there are few non-traditional pathways into teaching. Most secondary, primary,
and some early years ITE is delivered by one of seven universities. University
programmes of study are delivered face-to-face and/or online by distance. Early
years and primary programmes are typically 3 or 4 year undergraduate or one-
year graduate options; secondary programmes are typically a one-year graduate
programme. In one university, primary and secondary are offered as an integrated
4-year Bachelor of Education. In recent years, Master's level ITE has been offered
across several university programmes for both primary and secondary teaching.

The Teaching Council of Aotearoa New Zealand (formerly the Education
Council) approves, monitors, and reviews all ITE programmes with respect to
stated expectations related to graduates' readiness to teach. ITE programmes
all involve a mix of education and content-area coursework, taught completely
within educational faculties or in the case of some of the Māori medium courses
within Māori institutes, and are combined with regular field-placement experiences. Nine providers offer ten Māori-medium and bilingual ITE programmes and
one offers Pacific-based programmes (integrating Pacific pedagogy and practice)
in early childhood education (ECE) and primary education.

Despite concerns regarding equitable funding provisions for Māori-medium
teacher education, there is now a pathway into Māori-medium teaching where
20 years ago none existed, thus giving some recognition to the needs of Māori-
medium schooling. Expectations for Māori-medium immersion programmes are
that more than 80% of the content is taught in te reo Māori (more than 50%
for bilingual programmes) and that preservice teachers are provided with Māori-
medium teaching and learning experiences. With all curriculum papers based on
the Māori-medium school curriculum, the central focus is the sustainability of the

indigenous language te reo Māori (Stewart et al., 2018). Practicum placements are undertaken in Māori-medium kura (*schools*), and all student assignments are expected to be written in te reo Māori. However, efforts at the revitalization of te reo Māori within Aotearoa New Zealand are relatively recent. Consequently, many preservice teachers within Māori-medium teacher education lack the required levels of competency and confidence in te reo Māori to succeed in their ITE programmes and beyond in their schools (Lee-Morgan et al., 2019).

Equity-Focused ITE

The compulsory schooling sector in Aotearoa New Zealand is in a process of potential transformation. The Ministry of Education's (2018b) stated purpose is to "shape an education system that delivers equitable and excellent outcomes." Recommendations from the independent taskforce report, *Our Schooling Futures: Stronger Together* (TSIT, 2018), signal the need to "create innovative and flexible pathways into teaching … combined with opportunities that provide both broad and deep preparation for a range of contributions and advancement across the schooling system as a whole" (p. 36). Noting the availability of some excellent teaching, the TSIT authors articulate a vision of schooling in Aotearoa New Zealand as a continuously learning system aimed at reducing what they identify as considerable variability in teaching quality. Given the central influence of teaching quality on student success, they argue that a more coherent and connected policy for the recruitment, preparation, and induction of new teachers is needed, and that the responsibility for this should be shared amongst government agencies, ITE providers, and schools. In matters of ITE provisions, the taskforce concludes:

> We need to encourage greater teacher and teacher educator workforce diversity. We need to reduce the variability in the quality of graduates leaving teacher education. Finally, we need to increase the retention rate of beginning teachers/Kaiako. (TSIT, 2018, p. 87)

An important recommendation regarding preparation of teachers focuses on the need for ITE programmes to proactively support preservice teachers to develop pedagogies that strengthen inclusion/social justice/cultural responsiveness. Accountability is to the forefront of the recent *ITE Programme Approval, Monitoring and Review Requirements* (Teaching Council of Aotearoa New Zealand [TCANZ]), 2019), which include expectations that ITE providers can, for example, provide confidence that "assessments across the programme capture the student teacher's capability to work effectively with diverse learners, in multiple settings" (p. 28).

In meeting the challenge of supporting preservice teachers to develop culturally responsive capabilities and ensuring beginning teachers are able to act as change agents, ITE institutions are actively engaged in research-informed initiatives focused on equity. For example, the project, *Rethinking Initial Teacher Edu-*

cation for Equity (RITE), investigated preservice teachers' perceptions of how their ITE programme that was "specifically designed to put equity front and centre" (Grudnoff et al., 2016, p. 451) prepared them for teaching in low SES communities. A framework of six *facets of practice for equity* was embedded within the programme design:

1. Selecting worthwhile content and designing and implementing learning opportunities aligned to valued learning outcomes;
2. Connecting to students' lives and experiences;
3. Creating learning-focused, respectful, and supportive learning environments;
4. Using evidence to scaffold learning and improve teaching;
5. Adopting an inquiry stance and taking responsibility for professional engagement and learning; and
6. Recognizing and challenging classroom, school, and societal practices that reproduce inequity.

Embodying the facets in the content, teaching, and assessment of each course "prompted a move away from the traditional curriculum areas of primary school teaching (and ITE)" (p. 456); for example, "mathematics and literacy were intentionally integrated and taught by cross-discipline teams in three courses across the programme" (p. 456). Findings from the project are positive, and recent developments include the trial of a self-report, *Teaching Equity Enactment Scenario Scale* (TEES) (Chang et al., 2019). Trials suggest that TEES has potential to be used as one ITE outcome indicator and to "inform understandings of how to prepare and support teachers to enact teaching practice that responds to diversity, challenges education inequities, and promotes social justice" (p. 81).

Other major research projects concern disruption to the high levels of deficit teacher expectations, most notably associated with Māori and Pacific learners (e.g., Hunter & Hunter, 2017; Turner et al., 2015). Although programmatic research interventions focused on strategies and practice of high-expectation teachers are situated in schools (e.g., Peterson et al., 2016), the successful adaptations to teacher practice, such as shifts towards flexible grouping, enhancement of the class climate, culturally responsive differentiation, and supporting students' goal setting, have informed ITE programme development. Later in the chapter, we discuss the impact of one such professional learning intervention on mathematics education, *Developing Mathematics Inquiry Communities* (Hunter et al., 2018).

Professional Teaching Experiences (Practica)

ITE programmes are required to "explicitly model the principles and practices of effective and adaptive teaching" across all courses (TCANZ, 2019, p. 22) with an expectation that preservice teachers have opportunities to experience school-based pedagogical approaches being articulated within courses. Furthermore, providers are expected to ensure that "every aspect of the ITE programme is integrated, so that there is not a sense of 'theory' and 'practice' being enacted separately in different settings" (p. 23). Supported by school-based mentors (Associate Teachers), practicum provides an opportunity for preservice teachers to develop and demonstrate capabilities and dispositions related to teaching standards (discussed below). A minimum of 80 days of classroom-based professional experience (practicum) placements is required, or 120 days for most programmes of three years or longer. In many cases, this constitutes more than 40% of the total teaching weeks of the programme. In addition, some programmes provide course-related contact. For example, in the *Learning the Work of Ambitious Mathematics Teaching* project, discussed later in this chapter, the mathematics methods course includes teaching instructional activities within university–based rehearsals and with small groups of students in a partner school.

Teaching Standards

Aligned with the teaching-as-inquiry model promoted in the New Zealand Curriculum (MoE, 2007), considerable debate has been directed to revisions of graduating standards for ITE programmes, with the focus on practice and inquiry. For example, Sinnema et al. (2017) proposed a model of *Teaching for Better Learning* that "foregrounds the salience of teachers' own situations and the active nature of teachers' practices in a way that integrates practices with relevant theory" (p. 9). Although not adopted per se, teaching-as-inquiry is prominent in the 2019 programme design approval process, with expectations that ITE programmes promote the necessary skills and knowledge for preservice teachers to "undertake research into their own practices and to make evidence-informed decisions about what is best for their learners" (TCANZ, 2019, p. 17).

Currently, graduation from an ITE programme requires preservice teachers to meet, in a supported environment, the Teaching Council's *Standards for the Teaching Profession* (Education Council, 2017). The six standards comprise overarching statements describing the characteristics of effective teaching in the context of Aotearoa New Zealand. They are designed to promote high-quality teaching with respect to:

- Demonstrating commitment to Te Tiriti o Waitangi (*Treaty of Waitangi*) partnership;
- Using collaborative, inquiry-based professional learning to positively impact learning;
- Developing and maintaining learning-focused professional relationships;
- Developing a learning-focused culture;
- Designing for learning based on knowledge of curriculum, pedagogy, and learners; and
- Engaging in teaching in knowledgeable and adaptive ways.

From 2019, the Teaching Council will implement a new Assessment Framework by which ITE providers assess the extent to which student teachers meet the standards (TCANZ, 2019).

Induction

During their first two years of teaching, new teachers undertake a programme of induction and mentoring with the support of an experienced mentor teacher. Guidelines for induction and mentoring programmes, underpinned by a state funded reduced teaching allocation, are designed to promote an educative mentoring relationship involving co-constructed learning conversations focused on professional practice and development (Education Council, 2016). Teachers collate evidence of their professional development towards meeting the standards outlined above. After two years, they can become fully certificated teachers with the support of an attestation from their employing principal. Even though there is consensus that the national mandated programme is of high quality, in practice, there is evidence of differential spaces and opportunities for teacher learning, particularly for change-of-career teachers or for new teachers who are in non-permanent positions (Anthony et al., 2011; TSIT, 2018).

MATHEMATICS EDUCATION

Mathematics education receives explicit attention in early-years curriculum documentation: *Te Whāriki. He Whāriki Mātauranga mō ngā Mokopuna o Aotearoa: Early childhood curriculum* (MoE, 2017b). For instance, learning outcomes include "recognizing mathematical symbols and concepts and using them with enjoyment, meaning and purpose" (p. 25). Assessment of early-years mathematics learning involves noticing, recognizing, and responding to "children participating in mathematical practices–exploring relationships and using patterns in quantities, space, and time" (MoE, 2009, p. 2), and includes "noticing cultural and local conventions to do with ways of classifying and describing patterns and relationships, using ideas like number, shape, space, time, and distance" (p. 5). The curriculum draws on the metaphor of te kākano (*the seed*) for describing the range of activities for developing mathematical tools and symbol systems in a bicultural environment. The metaphor represents the child as te kākano embedded in a context. The range of mathematical purposes and tools that develop is influenced by the "fertilizer" or "soil" that surrounds te kākano. Influences include teacher knowledge and pedagogy as well as whānau (*family*) knowledge and resources, all of which interact with the child's interests to privilege mathematical domains.

For the school sector, Years 1–13, mathematics is specified as part of a whole school curriculum document: *The New Zealand Curriculum* (MoE, 2007). Mathematics education is designated as *Mathematics and Statistics,* structured around the three broad strands of number and algebra, geometry and measurement, and statistics. The first Māori-medium mathematics curriculum, *Pāngarau* (MoE,

1994), was introduced as a parallel to the English-medium mathematics curriculum. This was followed by the publication of *Te Marautanga o Aotearoa* (MoE 2008a), then revised in 2017 (MoE 2017a), which comprises nine learning areas, including pāngarau (*mathematics*). Rather than a direct translation of the New Zealand curriculum, *Te Marautanga o Aotearoa* is a curriculum based on kaupapa Māori (*Māori philosophies*) (Scoop Media, 2007). The learning area pāngarau is built on mathematically rich traditional activities including building, navigation, and food production and requires that learning be situated in Māori contexts and familiar experiences (MoE, 2017a).

Indigenous groups' endeavors to elaborate their own indigenous languages to teach and learn mathematics were closely linked to developments of the initial Māori-medium pāngarau (*mathematics*) curriculum. For instance, in developing a Māori mathematics register, there was early recognition that the standardization of a corpus of terms by itself was not sufficient (Barton et al., 1998). The lexical expansion required new and/or different ways of expressing mathematical ideas and concepts in te reo Māori. In practice, the development and implementation of the pāngarau curriculum has been significantly more resourced than other Māori-medium curriculum areas because of the role of mathematics as a government educational priority and also, relatedly, because of its high-stakes positioning in the education system (Trinick, 2019). Moreover, the pāngarau curriculum, in legitimizing the teaching of mathematics in te reo Māori, meant that mathematics came to have a pivotal role in supporting the reclamation of te reo Māori (Parra & Trinick, 2018).

However, the pāngarau curriculum is not without its critics. Tweed (2015) claims that attempts to relate mātauranga (*Māori knowledge*) and pāngarau have resulted, in some instances, in curriculum resources that prompt "symbolic violence to mātauranga [knowledge] and traditional practices by reconstructing traditional practices as mathematical practices, and pre-colonization Māori people as mathematicians" (p. 257). Efforts to contextualize pāngarau within traditional indigenous practices serve to present an impoverished view of mātauranga. Rather than be content to work with the existing mathematics curriculum, with reference to teaching and learning kaupapa (*policy, purpose*) and mātauranga mathematics, Te Maro (2018) calls for consideration and critique of "the possibility for maths education to be developed by informed and conscientized communities," those that are aware of the social and political conditions that impact on their ability to make emancipatory decisions and take transformative action through their own contexts, languages, and practices (p. 245).

Twenty years on from Māori curriculum developments, Trinick (2019) claims that evidence that the "kura sector is still struggling to incorporate Māori mathematical practices into school mathematics" (p. 10) suggests that language rights by themselves do not guarantee emancipatory forms of education. However, Te Maro (2018) takes pains to point out that the instances of problematic implementation of mathematics education in kura are "not because Māori and kura communities cannot do mathematics, do not want to engage with it, or do not find value in it. It is because maths educational outcomes are afforded higher status to kaupapa" (p. 259). According to Te Maro, in kaupapa English schools (schools that include Māori immersion options), this concern is not so visible "because kaupapa English schools are not having to simultaneously fight for the survival of their language and culture" (p. 259). To enable a seamless interaction from home to kura and back, Te Maro calls for the development of new programmes that would take cognizance of the interfaces between Māori knowledge and mathematics knowledge, where time, space, and activity would be used as sites of focus to determine that knowledge systems receive equitable privileging.

In addition to the mandated curriculum documents, there have been two Ministry of Education initiatives that have had significant impact on the teaching/learning of mathematics. Firstly, the *Numeracy Development Project,* a nationwide initiative implemented from 2000 to 2009 in both English and Māori-medium primary and lower secondary levels, advocated maximizing achievement gains through teaching students as much as possible in *ability* groups. In contrast, mixed ability grouping is portrayed as "contributing to the development of key competencies, such as relating to others, self-management, and belonging" (MoE, 2008b, p. 12). An evaluation report, *Mathematics in Years 4—8* (Education Review Office, 2013), suggested that, rather than accelerate the progress of struggling learners, the prevalence of ability grouping in mathematics has served to exacerbate existing disparities in achievement levels. Persistent underachievement and inequities were confirmed in the *National Monitoring Study of Student Achievement* (University of Otago & NZCER, 2014), where only 41% of all students, 11% of Pacific, and 26% of Maori were reported to be achieving curriculum standards in year 8.

A second Ministry of Education initiative during this period (which ceased from 2018) was the adoption of national standards for mathematics as well as for reading and writing. The revelation that some schools, in efforts to meet public reporting expectations, were targeting support, with most effort being put into those children who were just *below* and who could be shifted to *at* the standard (Thrupp & White, 2013), is an example of educational triage that, again, was disadvantageous to Māori and Pacific students.

Although there is emergent policy recognition of the need to provide more equitable participation and learning opportunities in mathematics classrooms (Pomeroy, 2018), there remains widespread uncertainty from mathematics teachers as to how to implement changes, most notably in grouping practices (Anthony & Hunter, 2017) and culturally responsive practices within mathematics classrooms (Hunter & Hunter, 2017). A large-scale university-led initiative, *The Developing Mathematical Inquiry Communities* (Hunter et al., 2018), has found that transformative change in grouping practices and associated equitable pedagogies requires whole-school, long-term, professional development support.

MATHEMATICS EDUCATION
WITHIN INITIAL TEACHER EDUCATION

In this section we focus on the recruitment of and mathematics education courses taken by prospective teachers. To illustrate current research-based reforms within initial teacher education programmes, we provide a case study on learning the work of ambitious mathematics teaching in ITE and associated developments in schools.

Recruitment

With regards to mathematics teaching, Aotearoa New Zealand experiences a long-standing recruitment issue on two fronts. Firstly, at the secondary level, mathematics is one of several curriculum areas that fail to attract sufficient teachers to fill teacher supply shortages and address declining student achievement. The Ministry of Education's offer of scholarships, including tuition fees and a stipend, has failed to increase enrolment numbers (~160) of prospective secondary teachers of mathematics and related STEM subjects from 2011 to 2017 (see MoE, 2019b).

Secondly, with pressure for ITE providers to offer more spaces for teacher candidates, there are raised tensions around entry level mathematics content knowledge, particularly for non-specialist early years and primary teachers. For some time, studies of teacher candidates' entry mathematical content knowledge have highlighted that compulsory papers in mathematics methods would be insufficient to address mathematics knowledge competency (Linsell & Anakin, 2012; Young-Loveridge et al., 2012). As recently as 2018, Ingram, Linsell, and Offen assessed 40% of their 83 entry primary teacher candidates as not meeting the required standard for content knowledge and exhibiting a negative relationship with mathematics. However, Murphy (2012) questioned the assumption that sound mathematics knowledge is necessarily transformed into effective teaching. Researching efforts to support preservice teachers' use of enquiry-based approaches, Murphy explored how the mathematics knowledge of four primary preservice teachers influenced their teaching of the topic of area. She found that the three preservice teachers who were confident in their mathematics were more predisposed to give sufficient explanations in relation to what they knew. This contrasted with the behaviour of the one preservice teacher, who to compensate for limited knowledge and confidence, self-researched the topic through pedagogical texts, an action that appeared to add "further exposure to teaching approaches that were likely to suggest an enquiry-based approach" (p. 203). Although Murphy's study affirmed the need for mathematics methods courses to focus on much more than improvements in mathematical content knowledge, collectively mathematics educators and the Teaching Council contend that pre-assessment of teacher candidates' mathematics knowledge is crucial, both as a threshold filter for recruitment and for the identification of specific areas that need to be supported within ITE.

Current regulations for programme approval require prospective teachers' numeracy skills to be assessed by ITE providers prior to admission and students are required to demonstrate numeracy skills that support their abilities to "work with the relevant curriculum" (TCANZ, 2019). For secondary teaching, ITE programmes require completion of undergraduate content-based courses appropriate for teaching two subjects in the secondary school curriculum. To specialize in mathematics, teaching candidates need to have completed undergraduate mathematics study (taught by mathematicians) at 200-level (intermediate undergraduate mathematics and/or statistics courses) for those aiming to teach junior secondary, and 300-level (advanced undergraduate mathematics and/or statistics courses) for those aiming to teach senior secondary.

Within the university-based early years and primary ITE programmes, prospective teachers are required to meet university-wide numeracy and literacy requirements.[2] As noted above, concerns that these standards are insufficient has led to ITE providers including a variety of additional mathematics tests within the recruitment process and/or providing additional intervention support during the ITE programme itself. Current debates about pre and post mathematics content knowledge standards (Education Council, 2016; Linsell & Anakin, 2013) have led to calls for a common assessment. However, as yet the details of what is to be assessed, for whom and at what standard, are largely unspecified (TCANZ, 2019).

Mathematics Education Courses

Mathematics education in the form of specialized mathematics methods courses is a key part in all ITE programmes, inclusive of early years to secondary qualifications. Methods courses are taught exclusively within ITE by research active mathematics teacher educators who typically have come from a practitioner background. In the last decade, several university ITE programmes have experienced a shift from the 3- or 4-year undergraduate mode to one-year postgraduate options. This shift has resulted in a decrease in the number of available mathematics education courses. Moreover, for those universities that have accepted prioritized Ministry of Education funding for master's-level ITE programmes, the requirement for closer collaboration between partner schools and universities, through extended school placement time, has meant further reductions in time available for mathematics methods courses. Programmes have coped by reducing the number of hours for mathematics education; in some programmes, dedicated mathematics methods courses have been integrated into general curriculum courses. Within the clinical-based master's model, anecdotal experiences suggest that the increased reliance on the uncertain expertise of mathematics teachers in the field (Ward & Thomas, 2007) has, in some instances, been a source of concern, both in terms of supporting the growth of preservice teachers' statistical and mathematics

---

[2] https://www.nzqa.govt.nz/qualifications-standards/awards/university-entrance/

content knowledge (Pfannkuch & Ben-Zvi, 2011) and in access to mathematics classroom experiences that model ambitious mathematics pedagogies.

Teacher educators in Aotearoa New Zealand take seriously the agenda of reform in mathematics classrooms, advocating a vision of ambitious teaching (Lampert et al., 2013) where teachers engage students in challenging tasks and collaborative inquiry, and observe and listen as students work so they can provide appropriate levels of support to diverse learners. Such ambitious mathematics teaching involves "skilled ways of eliciting and responding to each and every student in the class so that they learn worthwhile mathematics and come to view themselves as competent mathematicians" (Anthony, Hunter, Hunter, et al., 2015).

To the forefront of supporting preservice teachers to learn the work of ambitious mathematics teaching, mathematics education courses have increasingly focused on the development of an inquiry stance (Aitken et al., 2013) that supports growth in professional noticing, culturally responsive pedagogies, and specialized content knowledge. To illustrate adaptations to pedagogies and impacts of this work, we present a case study of reform enacted across two university ITE programmes (Anthony, Hunter, Anderson, et al., 2015) inclusive of a feedback/feed-forward loop associated with the *Developing Mathematical Inquiry Communities* (Hunter et al., 2018), which is a school-based research and professional learning project. Before presenting the case study, we briefly overview ITE mathematics education courses within Aotearoa New Zealand.

Although the means and opportunities to learn the work of ambitious mathematics teaching varies between programmes and ITE providers, a review of prescriptions for mathematics education methods papers reveals considerable commonality. Aligned with the *New Zealand Curriculum* (MoE, 2007), content knowledge includes reference to statistical and mathematics content knowledge and thinking—for example, "Develop knowledge and understanding of the nature of mathematics and statistics…what is meant by thinking mathematically and statistically" (University of Auckland, 2019). In terms of espoused pedagogy, a noteworthy commonality across mathematics education course prescriptions is the framing of sociocultural theories of learning with explicit reference to concepts of engagement in learning communities—for example, "…to explore how teachers and learners function and engage in learning communities, including the classroom, the school and the wider social and cultural community" (Eastern Institute of Technology, 2019).

In supporting preservice teachers' capability to "design and plan culturally responsive, evidence-based approaches that reflect the local community and Te Tiriti o Waitangi partnership in New Zealand" (Education Council, 2017, p. 20), there is increasing recognition that mathematics methods courses are legitimate sites for preservice teachers to learn how to implement culturally responsive pedagogies and plan for diversity and inclusion (Averill & McRae, 2019). Within mathematics methods courses, this inclusion signals explicit recognition of students' whānau (*family*) and wider communities—for example, "Recognize, support and plan to extend the mathematics and literacy understandings that children bring to early childhood contexts" (Massey University, 2019a) and expectations that preservice teachers will demonstrate the "ability to incorporate te reo Māori me ngā tikanga Māori (*Māori language and customs*) in their relevant curriculum areas" (University of Otago, 2019). With reference to policy documents, courses have moved to press preservice teachers to make explicit reference to competencies outlined in *Tātaiako: Cultural Competencies for Teachers of Māori Learners* (MoE, 2011) and *Tapasā: Cultural Competencies Framework for Teachers of Pacific Learners* (MoE, 2019c) within their planning documentation—for example, "Evaluate mathematics resource material for use in primary classrooms including bicultural and multicultural settings" (Massey University, 2019b). In a researched initiative, Wilson et al. (2017) modelled ways that a bicultural perspective could be included in mathematics methods courses through the inclusion of active partnership between learners, Māori language, as well as Māori pedagogies, contexts, beliefs, philosophies, protocols, and values. Wanting to design learning opportunities that offered more than cosmetic and/or tokenistic changes to task design, their approach was to present mathematics as being "of and from our everyday human realms, sitting right there in our culture" where we "talk it, argue it, and describe it in more than one language, and in many contexts" (Averill et al., 2010, p. 176). Specifically, they required planning, both within their courses and during teaching experience, to make explicit references to competencies of *Tātaiako: Cultural Competencies for Teachers of Māori Learners* (MoE, 2011). Offering further strategies to model culturally responsive pedagogies, Averill (2018) describes the use of song (e.g., action songs about measurement concepts involving measurement terms in te reo Māori), oral storytelling, and metaphor (e.g., through a practical culturally-based activity of making a physical representation of their learning in a tukutuku-style panel, which is a traditional woven design).

Across ITE programmes, understanding the role of assessment reflects the Ministry of Education's current focus on evidenced-based data accountability in schools (Education Review Office, 2017) and a focus on differentiated instruction (Anthony et al., 2019), as exemplified by stated learning outcomes—for example, "Demonstrate proficiency in a range of teaching and learning approaches for differentiated classes in mathematics and statistics and use evidence-based practice to establish learning needs" (Victoria University of Wellington, 2019). However as noted earlier, the current predicament of varied grouping practices within schools—for example, ability grouping as advocated in the New Zealand Numeracy project (MoE, 2008b) or strength-based grouping as advocated in the *Developing Mathematics Inquiry Communities* school-based professional learning project (Hunter et al., 2018)—make this a contested space in ITE programmes and schools.

To address the concerns about primary teacher candidates' entry level mathematics knowledge, several ITE programmes include voluntary or required intervention courses to support preservice teachers' mathematics knowledge development. In contrast to an add-on model, the *Growing Mathematics Teachers* project (Ingram et al., 2018) builds on a growth perspective (Dweck, 2016) that utilizes an adaptive assessment at the beginning of their ITE programme with the expectation that preservice teachers will "decide their own next steps for learning" (p. 47). Throughout their study, preservice teachers choose to support their learning through access to content-based mathematics websites (e.g., MoE, 2019a) or through study of an additional mathematics content course. Second year courses also provide explicit opportunities to explore one's relationship with mathematics through metaphor exercises and personal journey graphs. Ingram and colleagues (2018) contend that the success of the intervention was in part due to the interactions between mathematical content knowledge, affective elements, and the growth perspective, supported within a longitudinal timeframe.

Case Study: Learning the Work of Ambitious Mathematics Teaching in ITE and Schools

In looking to respond to the challenges of learning the work of ambitious teaching, several mathematics ITE programmes across Aotearoa New Zealand (e.g., Anthony, Hunter, Anderson et al., 2015; Bailey & Taylor, 2015; Murphy, 2016) have embraced the international trend towards practice-based reforms that "view teaching not only as a resource for learning to teach but as a central element of learning to teach" (McDonald et al., 2014, p. 500). To illustrate the nature of the reforms and the impact for learning, for preservice teachers, for teacher educators, and for beginning teachers, we draw on the *Learning the Work of Ambitious Mathematics Teaching* (LAMT) project (Anthony, Hunter, Anderson et al., 2015). With regard to teacher educator learning and ongoing programme development, we also discuss a critical and additional source of learning occasioned by teacher educators'/researchers' parallel involvement in the development of the school-based professional learning intervention, *Developing Mathematical Inquiry Communities* (DMIC) (Hunter et al., 2018).

LAMT, a 3-year project jointly implemented in two major ITE institutions (the authors' and Victoria University of Wellington), involved trialing new ways to support preservice teachers to not only *think* like teachers but also to put what they know into action. Modeled on the seminal work of the *Learning In, From, and For Teaching Practice* project (Lampert et al., 2013), the design-based project drew on the practice-based pedagogical framework of Grossman, Hammerness, and McDonald (2009) involving representation of teaching (e.g., modelling), decomposition of practices (e.g., focus on core/high-leverage practices), and approximation of practice (e.g., rehearsals). Our mathematics methods courses, for both primary and secondary education, were adapted to incorporate learning experiences that began with preservice teachers observing teacher-educators modelling purposefully designed Instructional Activities (IAs), followed by planning and teaching IAs both in methods courses and school-based settings (see Figure 6.1). Adaptable to multiple grade levels, the IAs, activities such as choral count-

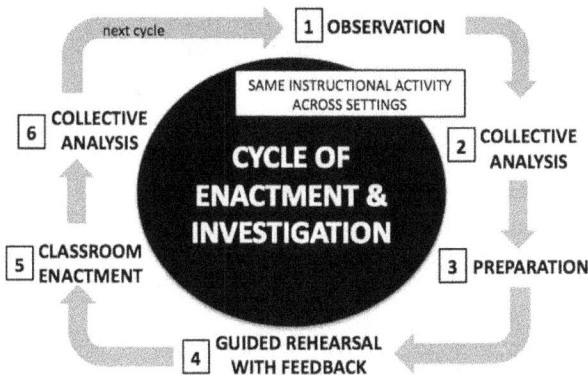

FIGURE 6.1.   LTP model of cycles of enactment and investigation from Lampert et al. (2013). (Reprinted with permission of Sage Publications)

ing, strings, and quick images (see Kazemi et al., 2016), act as containers of core practices, pedagogical tools, and principles of high-quality, ambitious mathematics teaching.

We knew that occasioning shifts in preservice teachers' professional vision about mathematics teaching would require that our preservice teachers become "inquiring professionals who are focused on better learning for themselves and their students" (Aitken et al., 2013, p. 30). If done well, we envisaged that the cycle of enactment and investigation would provide real-time, practice-based opportunities to learn the interactive work of mathematics teaching, where teachers act deliberately and responsibly in ways that support diverse students to become powerful learners.

Central to this learning experience, and a major change for preservice teachers' opportunities to learn, was the incorporation of public rehearsals of IAs across a range of primary and secondary level university methods courses. The rehearsal phase is where nominated preservice teachers teach a pre-planned IA to a group of their peers acting as student learners, accompanied by in-the-moment coaching provided by the teacher educator. Designed to support collaborative inquiry, these deliberate approximations of practice provide spaces for preservice teachers to "open up their instructional decisions to one another and their instructor" (Kazemi et al., 2016, p. 20).

The role of the coach, so integral to the success of the rehearsal phase, required that we as teacher educators learned to engage in practice-based pedagogies. Our learning journey required that we press rehearsing preservice teachers and their peers to think deeply in an evidenced-based way "about specific teaching (re)actions in relation to opportunities for each student's participation and potential and actual learning" (Anthony et al., 2018, p. 8). We found that the very process of facilitating professional noticing of teaching moves and consequent student outcomes also served to strengthen our understanding of preservice teachers' knowledge and beliefs about learning and teaching mathematics. For example, early rehearsals surfaced preservice teachers' disquiet about whether the representation of 3×4 should be 3 groups of 4 or 4 groups of 3, or both? Moreover, participation in the role of teacher and/or learner exposed preservice teachers' awkwardness in providing and listening to mathematical explanations.

Aligned to the course objectives for our mathematics methods courses, the cycle of enactment and investigation (see Figure 6.1) provided rich opportunities for our preservice teachers to learn the work of planning, noticing, and responding to learner's mathematical thinking; to build communities of learners; to experience modelling and development of cultural competencies; and to strengthen their mathematical content knowledge. In the spirit of collaborative inquiry, we also sought input of student voice though surveys and post-rehearsal video-stimulated recall interviews (Rawlins et al., 2019). Extending beyond an openness to hearing preservice teachers' perspectives, according them partnership status in the research empowered them to take an active role in shaping their learning. Findings affirmed that rehearsals formed a crucial part of our preservice teachers' learning trajectory, including enhancement of feelings of efficacy and development of an inquiry stance. In terms of the rehearsal process, of note was preservice teachers valuing the dual roles of being either the rehearsing teacher or being one of the student cohort.

As discussed by the preservice teachers, the rehearsal process, through opportunities to question, to discuss, and retry teaching moves, supported a trajectory of learning. Drawing initially on decomposition of teacher educator modelling of practice and the sharing of expertise within collaborative planning, early rehearsals tended to focus on skill-based practices (e.g., voice projection and turn taking). As the process of rehearsal became familiar and linked more closely with small-group teaching within schools, the focus shifted towards high-leverage instances (Stockero et al., 2017) concerning making students' mathematical thinking and reasoning visible and accessible to peers. To illustrate, Anthony (2018) provides an episode from a choral count rehearsal (see Figure 6.2), in which the rehearsing teacher's eliciting process is extended from having peers engage with a particular response towards using the response as a building block to further the student discussion. We enter the rehearsal immediately after the rehearsing teacher (RT) records Robert's suggested pattern of "55 being added to each number" (pointing to diagonal pairs).

In this instance, the coach's suggested teacher move prompted RT to explore an alternative way to support students to engage with their peers' reasoning. Moreover, the coach's feedback made reference to impact in terms of how the learner was scaffolded to engage with the structural nature of the pattern. As such, it served to draw attention to linking the teacher move to the opportunity to learn. The aim of this explicitness was to enable the preservice teachers to access essen-

RT:	That's good. Does anyone have another pattern?
Coach:	Pause. That's quite a complex idea and it might be one which you want to throw back to them and say does everyone agree? Like, "Let's look at what Robert said; he said that they increase by 55. Do you agree, why or why not"?
RT:	Right, I would like you all to have a think about what Robert just shared with us because that is quite a complex idea, and think about what Cath said at the start about how she adds five, and somebody else said that when we are going down we are adding five tens, so think about that, adding five [pause]. *Oh I am giving it away aren't I?* Have a chat to your neighbor about how that works.

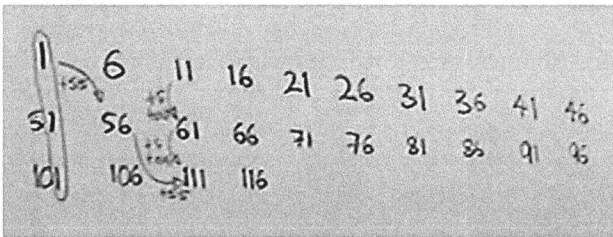

After the group of rehearsing students had talked for a few minutes, RT asked them to share their ideas:

Megan:	If you go across it is plus 5 and then going down is five tens so 5 times 10 is 50 so the 5 plus the 50 is 55 [RT notates the explanation].
RT:	So that way is the same as those two? Is that what you are saying [notating the explanation with arrows]?
Megan:	Yes you can add them together.
RT:	Great.
Coach:	Pause. You know you said *I am kind of giving it away* but what I think RT did was you really structured it so they could work out why that pattern was. If you had just said just look at it, with Year Fours they may not have seen it. You didn't say what you need to do is…, but you said look at that idea, and look at that idea, and that gave a foundation for them to then see that and use that, so that was a good thing to do.

FIGURE 6.2. Choral count episode adapted from Anthony (2018)

tial processes in their practice and become students of their students and learners of their own practice.

When "teachers explore their students' learning, they adopt a different stance, placing themselves in the role of learners" (Hadar & Brody, 2016, p. 102), a stance that is central to developing cultural competencies. Averill et al. (2015) illustrate how preservice community interactions around teacher educator modelling of the IAs (the representation phase) and in-the-moment coaching can provide oppor-

132 • ANTHONY & EDEN

tunities to sensitize preservice teachers towards culturally responsive teaching practices associated with *Tātaiako: Cultural Competencies for Teachers of Māori Learners* (MoE, 2011). For example, the opportunities to learn how to support effective interactions, manage co-construction, and use cooperative learner-focused activities modelled the concept of wānanga (*the participation with learners in robust dialogue for the benefit of learners*) and ako (*taking responsibility for one's own learning and the learning of others*). Moreover, the press within the rehearsal process to attend to who the students are and how they are positioned within the learning aligns well with the cultural competency of tangata whenuatanga (*affirming Māori learners as Māori*).

Building cultural competencies is part and parcel of developing culturally responsive practices, and responsive pedagogy involves dialogue where "risk taking is encouraged, where there is no shame in being a 'not knower' and where it is understood that everyone brings with them knowledge, ways of knowing, and experiences of value to share" (Berryman et al., 2018, p. 7). Within the cycle of enactment and investigation, teaching that values and respects each student's mathematical thinking and places that thinking at the centre of decision making was foregrounded. Aware that learning to listen effectively and respond to the multiplicity of features specific to students' thinking "is surprisingly hard work" (Empson & Jacobs, 2008, p. 257), we deliberately selected IAs to support a trajectory of learning associated with noticing students' mathematical thinking (Anthony, Hunter, & Hunter, 2015b). We moved from elicitation of explanations, then responding towards building on and connecting students' thinking to the mathematical goals of the lessons. As student actors in the rehearsals engaged more frequently in risk taking and issues of working with errors, discussions around student participation and assignment of status surfaced within instructional practice decisions. Alongside this trajectory we noted our coaching moves shifted from more directive to be more inclusive of our learning community, thereby engaging all participants in richer, more wide-ranging reflections about teaching practices and impacts (Averill et al., 2016).

Preservice teachers' teaching IAs within school settings is an important part of the cycle of enactment and investigations. Working with small groups, preservice teachers collectively plan and (co)teach or act as peer observers for IAs with small groups of students, and engage in post-teaching reflective discussions within their methods courses. Repeated four times in a course, this activity is in addition to the programmatic school teaching experience. Anthony, Hunter, and Hunter (2015a) describe how this inquiry building cycle supported the development of adaptive expertise. The requirement to teach students in mixed achievement groups and the press to collaboratively evaluate their instructional practices in terms of students' participatory practices and learning outcomes positively impacted preservice teachers' shifts in focus from self to student, and supported the development of more complex understandings of teaching and learning. The repeated opportunities to engage with the same group of students were described by preservice teachers as an experimentation process informed by coursework readings and rehearsals. This process further supported their development of an ethic of care associated with "inquiring professionals who are focused on better learning for themselves and their students" (Aitken et al., 2013, p. 30), a hallmark of adaptive expertise.

The last year of the LAMT project involved following a group of beginning teachers. Although each beginning teacher journey was unique, we found repeated instances of beginning teachers acting as change agents within schools. For example, for Shelly, evidence of her increasing awareness to become her own teacher and utilization of the skills to learn from and for practice was apparent in her role as resource, in terms of changes in group activities, for her more senior colleagues (Anthony, Hunter, Hunter, & Duncan, 2015).

Of note, our work on practice-based pedagogies coincided with the expansion of the *Developing Mathematical Inquiry Communities* (DMIC) project involving professional learning and development within schools situated in low socio-economic communities. With several teacher educators/researchers in common to both projects, the DMIC project adapted the in-the-moment coaching experiences of LAMT for in-class mentoring. Termed *dynamic mentoring* (Hunter, Hunter, Bills, & Thompson, 2016), the mentors work alongside teachers to co-construct lessons in the moment and support teachers to critically reflect on and reshape classroom practices.

A direct consequence of teacher educators'/researchers' ongoing involvement in DMIC has been our continued learning, not just *about* the work of ambitious teaching, but engagement with the *practice of* ambitious teaching. Underpinned by a communication and participation framework tool (Hunter, 2008), teachers learn to adaptively and flexibly support students to engage in mathematical practices related to inquiry and the development of productive dispositions. Driven by socio-political concerns to provide equitable mathematics education, DMIC provides a living resource of ambitious teaching practice that we can continue to draw on in our ITE courses. In addition to our own experiences working in DMIC classrooms and with DMIC teachers, the resource includes video exemplars of culturally responsive practice in action, equitable grouping, collaborative planning, and classroom talk (see Massey University, 2017). Additionally, research focused specifically on teacher professional learning within the DMIC programme provides insights into *smart tools* that could be used within ITE, for example, Leach's (2019) smart planning tool for strength-based grouping which is a template "designed to support teachers in identifying individual students' strengths as criteria for grouping students for collaborative sense-making" (p. 428).

Measures of impact of the various ITE programmes within Aotearoa New Zealand in terms of the preparation of teachers who provide quality teaching and learning outcomes are difficult to determine (Ell et al., 2019). In determining impact, Grudnoff et al. (2016) argue that we must increasingly attend "to the complex and multi-layered contexts, school and policy environments in which student teachers learn to teach, and the larger structures of advantage and disadvantage that intersect with these" (p. 452). In proposing a model based on complexity theory, Ell et al. (2019) contend that we need to broaden our recognition of the scope of influence within education systems surrounding students. For example, we should include partnerships with schools, educator mentor growth, professional learning of school staff, and the production of educational research. As seen in the case study, working within DMIC, a collaborative partnership between teacher educators and school communities, has served to benefit students, teachers, and teacher educator learning.

GOING FORWARD

In Aotearoa New Zealand, a commitment to Te Tiriti o Waitangi and equity within mathematics classrooms calls for transformation of the way we teach mathematics (Hunter et al., 2018), and thus the way we prepare mathematics teachers. As noted by Averill (2018), the persistent inequities in mathematics learning opportunities within Aotearoa New Zealand "show that we must seek to [prepare teachers who can] make big waves to disrupt thinking to enhance and supplement commonly found mathematics teaching practices" (p. 21). In challenging the inequities in mathematics classrooms, Te Maro (2018) argues that our focus should not be on preparing teachers who can "get our tamariki (*children*) better at maths education." Rather, in response to inequitable power relations, cultural invisibility, forced cultural adaptation and enculturation, and attempts toward assimilation through deliberate policy, our approach should be "to conscientize us all to what maths education is for, who influences it, how, and why" (p. 235).

Going forward with an equity focus requires that we strengthen the emergent research focus involving the nature of knowledge/learning and curriculum values (Hill et al., 2019). Particularly, given concerns that mathematics education:

> is a system that narrows mathematics to a shadow of its true self. This is done by shrinking mathematics into forms that can support the normalizing, standardizing, and disciplining of all children. Further purpose for this type of maths education development is to satisfy societal (capitalist and neoliberal) priorities and purposes. (Te Maro, 2018. p. 235)

Challenging Eurocentric mathematics education will require greater efforts to attend to the Māori (Tweed, 2015) and Pasifika (Averill & Rimoni, 2019; Hunter, Hunter, Bills, Cheung et al., 2016) world views. For example, the study by Hawera and Taylor (2017) involving Māori-medium preservice teachers' experience with mentor teachers advised that more emphasis on the development and articulation of philosophy about mathematics learning would support preservice teachers' engagement in professional conversations in Māori immersion settings.

In addition to adapting curricula and programme structures that are equity centered, Cochran-Smith et al. (2016) argue that, in moving forward, teacher education needs to maintain a focus on reforming instruction. In focusing on both what we teach and how we teach, we must more deeply and explicitly tackle the "work that teachers need to learn to do in attending to power and status in their classroom" (Shaughnessy et al., 2019, p. 177). This will require we provide a safe space to prepare teachers who can negotiate practice "amid intersecting stories of traditional and reform movements" (Nolan & Walshaw, 2012, p. 345), who can challenge inequities, and who can attend to well-being alongside academic achievement (Berryman et al., 2018). To be generative in advancing our commitment to equity, we, as teacher educators, will need to take account of what preservice teachers know and can do, but also who they are—their histories, their emotions, and their well-being. Moreover, we will need to commit to our engagement to learn through partnerships, with preservice teachers and existing and new communities (e.g., Grudnoff et al., 2016), focused on communication and negotiation around the practice of teaching and teacher education.

REFERENCES

Aitken, G., Sinnema, C., & Meyer, F. (2013). *Initial teacher education outcomes: Standards for graduating teachers*. University of Auckland.

Anthony, G. (2018). Practice-based initial teacher education: Developing inquiring professionals. In G. Kaiser, H. Forgasz, M. Graven, A. Kuzniak, E. Simmt, & B. Xu (Eds.), *Invited lectures from the 13th International Congress on Mathematical Education* (pp. 1–18). Springer Open.

Anthony, G., Averill, R., & Drake, M. (2018). Occasioning teacher-educators' learning through practice-based teacher education. *Mathematics Teacher Education and Development, 20*(3), 4–19.

Anthony, G., Haigh, M., & Kane, R. (2011). The power of the 'object' to influence teacher induction outcomes. *Teaching and Teacher Education, 27*(5), 861–870.

Anthony, G., & Hunter, R. (2017). Grouping practices in New Zealand mathematics classrooms: Where are we at and where should we be? *New Zealand Journal of Educational Studies, 52*(1), 73–92.

Anthony, G., Hunter, R., Anderson, D., Averill, R., Drake, M., Hunter, J., & Rawlins, P. (2015). *Learning the work of ambitious mathematics teaching*. New Zealand Council of Educational Research.

Anthony, G., Hunter, J., & Hunter, R. (2015a). Prospective teachers' development of adaptive expertise. *Teaching and Teacher Education, 49*, 108–117.

Anthony, G., Hunter, J., & Hunter, R. (2015b). Supporting prospective teachers to notice students' mathematical thinking through rehearsal activities. *Mathematics Teacher Education and Development, 17*(2), 7–24.

Anthony, G., Hunter, J., & Hunter, R. (2019). Working towards equity in mathematics education: Is differentiation the answer? In B. White, M. Chinnappan, & S. Trenholm (Eds.), *Proceedings of the 42nd annual conference of the Mathematics Education Research Group of Australasia* (pp. 117–124). Mathematics Education Research Group of Australasia.

Anthony, G., Hunter, R., Hunter, J., & Duncan, S. (2015). How ambitious is "ambitious mathematics teaching"? *Set: Research Information for Teachers, 2*, 45–52.

Averill, R. (2012). Caring teaching practices in multiethnic mathematics classrooms: Attending to health and well-being. *Mathematics Education Research Journal, 24*(2), 105–128.

Averill, R. (2018). Examining historical pedagogies towards opening spaces for teaching all mathematics learners in culturally responsive ways. In J. Hunter, P. Perger, & L. Darragh (Eds.), *Proceedings of the 41st annual conference of the Mathematics Education Research Group of Australasia* (pp. 11–27). MERGA.

Averill, R., Anderson, D., & Drake, M. (2015). Developing culturally responsive teaching through professional noticing within teacher educator modelling. *Mathematics Teacher Education and Development, 17*(2), 64–83.

Averill, R., Drake, M., Anderson, D., & Anthony, G. (2016). The use of questions within in-the-moment coaching in initial mathematics teacher education: Enhancing participation, reflection, and co-construction in rehearsals of practice. *Asia-Pacific Journal of Teacher Education, 44*(5), 486–503.

Averill, R., & McRae, H. (2019). Culturally sustaining initial teacher education: Developing student teacher confidence and competence to teach indigenous learners. *The Educational Forum, 83*(3), 294–308.

Averill, R., & Rimoni, F. (2019). Policy for enhancing Pasifika learner achievement in New Zealand: Supports and challenges. *Linhas Críticas, 25*, 549–564.

Averill, R., Taiwhati, M., & Te Maro, P. (2010). Knowing and understanding each other's cultures. In R. Averill & R. Harvey (Eds.), *Teaching primary school mathematics and statistics: Evidence-based practice* (pp. 167–180). NZCER Press.

Bailey, J., & Taylor, M. (2015). Experiencing a mathematical problem-solving teaching approach: Opportunities to identify ambitious teaching practices. *Mathematics Teacher Education and Development, 17*(2), 111–124.

Barton, B., Fairhall, U., & Trinick, T. (1998). Tikanga reo tatai: Issues in the development of a Māori mathematics register. *For the Learning of Mathematics, 18*(1), 3–9.

Berryman, M., Lawrence, D., & Lamont, R. (2018). Cultural relationships for responsive pedagogy: A bicultural mana ōrite perspective. *Set, 1*, 3–10.

Caygill, R., Hanlar, V., & Singh, S. (2016). *TIMSS 2015: New Zealand Year 5 Maths results*. Comparative Education Research Unit, Ministry of Education.

Chamberlin, M., & Caygill, R. (2012). *Key finding from New Zealand's participation in the Progress in International Reading Literacy Study (PIRLS) and Trends in International Mathematics and Science Study (TIMSS) in 2010/11*. Ministry of Education.

Chang, W. C. C., Ludlow, L. H., Grudnoff, L., Ell, F., Haigh, M., Hill, M., & Cochran-Smith, M. (2019). Measuring the complexity of teaching practice for equity: Development of a scenario-format scale. *Teaching and Teacher Education, 82*, 69–85.

Cochran-Smith, M., Ell, F., Grudnoff, L., Haigh, M., Hill, M., & Ludlow, L. (2016). Initial teacher education: What does it take to put equity at the center? *Teaching and Teacher Education, 57*, 67–78.

Dweck, C. (2016). *Mindsets: The new psychology of success*. Random House USA Inc.

Eastern Institute of Technology. (2019). *Bachelor of Teaching (Primary) 2019*. https://www.eit.ac.nz/wp-content/uploads/programme-app-packs/int%20Bach%20Teaching%20(Primary).pdf

Education Council. (2016). *Strategic options for developing future oriented initial teacher education*. https://www.educationcouncil.org.nz/sites/default/files/Strategic%20options%20REVISED%2029 %20JUNEpdf.pdf

Education Council. (2017). *Our code our standards: Code of professional responsibility and standards for the teaching profession.* Author.

Education Review Office. (2013). *Mathematics in years 4 to 8: Developing a responsive curriculum.* Education Review Office.

Education Review Office. (2017). *Keeping children engaged and achieving in mathematics.* Education Review Office.

Ell, F., Haigh, M., Cochran-Smith, M., Grudnoff, L., Ludlow, L., & Hill, M. F. (2017). Mapping a complex system: What influences teacher learning during initial teacher education? *Asia-Pacific Journal of Teacher Education, 45*(4), 327–345.

Ell, F., Simpson, A., Mayer, D., McLean Davies, L., Clinton, J., & Dawson, G. (2019). Conceptualising the impact of initial teacher education. *The Australian Educational Researcher, 46*(1), 177–200.

Empson, S. B., & Jacobs, V. R. (2008). Learning to listen to children's mathematics. In D. Tirosh & T. Wood (Eds.), *Tools and processes in mathematics teacher education* (pp. 257–281). Sense Publishers.

Grossman, P., Hammerness, K., & McDonald, M. (2009). Redefining teaching, re-imagining teacher education. *Teachers and Teaching: Theory and Practice, 15*(2), 273–289.

Grudnoff, L., Haigh, M., Hill, M., Cochran-Smith, M., Ell, F., & Ludlow, L. (2016). Rethinking initial teacher education: Preparing teachers for schools in low socio-economic communities in New Zealand. *Journal of Education for Teaching, 42*(4), 451–467.

Hadar, L. L., & Brody, D. L. (2016). Talk about student learning: Promoting professional growth among teacher educators. *Teaching and Teacher Education, 59*, 101–114.

Hawera, N., & Taylor, M. (2017). "I learned quite a lot of the maths stuff now that I think about it": Maori medium students reflecting on their initial teacher education. *Australian Journal of Teacher Education, 32*(5), 87–100.

Hill, J., Hunter, J., & Hunter, R. (2019). What do Pāsifika students in New Zealand value most for their mathematics learning? In P. Clarkson, W. Seah, & J. Pang (Eds.), *Values and valuing in mathematics education* (pp. 103–114). Springer.

Hudson, B. (2019). *Petition of David Chester: Change the name of the country to Aotearoa—New Zealand and Petition of Danny Tahau Jobe: Referendum to include Aotearoa in the official name of New Zealand, report of the Governance and Administration Committee.* https://www.parliament.nz/en/pb/sc/reports/document/SCR_88256/petition-of-danny-tahau-jobe-referendum-to-include-aotearoa

Hunter, J., Hunter, R., Bills, T., Cheung, I., Hannant, B., Kritesh, K., & Lachaiya, R. (2016). Developing equity for Pāsifika learners within a New Zealand context: Attending to culture and values. *New Zealand Journal of Educational Studies, 51*(2), 197–225.

Hunter, R. (2008). Facilitating communities of mathematical inquiry. In M. Goos, R. Brown, & K. Makar (Eds.), *Proceedings of the 31st annual Mathematics Education Research Group of Australasia conference* (pp. 31–39). MERGA.

Hunter, R., & Hunter, J. (2017). Maintaining a cultural and mathematical identity while constructing a mathematical disposition as a Pāsifika learner. In E. McKinley & L. Tuhiwai Smith (Eds.), *Handbook of indigenous education* (pp. 1–19). Springer.

Hunter, R., Hunter, J., Anthony, G., & McChesney, K. (2018). Developing mathematical inquiry communities: Enacting culturally responsive, culturally sustaining, ambitious mathematics teaching. *Set, 2*, 25–32.

Hunter, R., Hunter, J., Bills, T., & Thompson, Z. (2016). Learning by leading: Dynamic mentoring to support culturally responsive mathematical inquiry communities. In B. White, M. Chinnappan, & S. Trenholm (Eds.), *Proceedings of the 38th annual conference of the Mathematics Education Research Group of Australasia* (pp. 59–73). MERGA.

Ingram, N., Linsell, C., & Offen, B. (2018). Growing mathematics teachers: Pre-service primary teachers' relationships with mathematics. *Mathematics Teacher Education and Development, 3,* 41–60.

Kazemi, E., Ghousseini, H., Cunard, A., & Turrou, A. C. (2016). Getting inside rehearsals: Insights from teacher educators to support work on complex practice. *Journal of Teacher Education, 67*(1), 18–31.

Lampert, M., Franke, M. L., Kazemi, E., Ghousseini, H., Turrou, A. C., Beasley, H., Cunard, A., & Crowe, K. (2013). Keeping it complex: Using rehearsals to support novice teacher learning of ambitious teaching. *Journal of Teacher Education, 64*(3), 226–243.

Leach, G. (2019). Strength-based grouping: A call for change. In B. White, M. Chinnappan, & S. Trenholm (Eds.), *Proceedings of the 42nd annual conference of the Mathematics Education Research Group of Australasia* (pp. 428–435). MERGA.

Lee-Morgan, J., Courtney, M., & Muller, M. (2019). New Zealand Māori-medium teacher education: An examination of students' academic confidence and preparedness. *Asia-Pacific Journal of Teacher Education, 47*(2), 137–151.

Linsell, C., & Anakin, M. (2012). Diagnostic assessment of pre-service teachers' mathematical content knowledge. *Mathematics Teacher Education and Development, 14*(2), 4–27.

Linsell, C., & Anakin, M. J. (2013). Foundation content knowledge: What do pre-service teachers need to know? In V. Steinle, L. Ball, & C. Bardini (Eds.), *Mathematics education: Yesterday, today and tomorrow* (Proceedings of the 36th annual conference of the Mathematics Education Research Group of Australasia) (pp. 442–449). MERGA.

Massey University. (2017). *DMIC: Developing mathematical inquiry communities.* https://www.cerme.nz/dmic

Massey University. (2019a). *265.473 Integrating early childhood curriculum: Mathematics and literacy.* https://www.massey.ac.nz/massey/learning/programme-course/course.cfm?course_code=265473&course_offering_id=1258943

Massey University. (2019b). *278.424 Mathematics teaching in the primary school.* https://www-massey-ac-nz.ezproxy.massey.ac.nz/massey/learning/programme-course/course.cfm?course_code=278424&course_offering_id=1266063

McDonald, M., Kazemi, E., Kelley-Petersen, M., Mikolasy, K., Thompson, J., Valencia, S. W., & Windschitl, M. (2014). Practice makes practice: Learning to teach in teacher education. *Peabody Journal of Education, 89*(4), 500–515.

Milne, A. (2016). Where am I in our schools' white spaces? Social justice for the learners we marginalise. *Middle Grades Review, 1*(3), 2.

Ministry of Education. (1994). *Pāngarau.* Learning Media.

Ministry of Education. (2007). *The New Zealand curriculum.* Learning Media.

Ministry of Education. (2008a). *Te Marautanga o Aotearoa.* Learning Media.

Ministry of Education. (2008b). *Numeracy professional development projects 2008, Book 3: Getting started.* Learning Media.

Ministry of Education. (2009). *Kei tua o te pae assessment for learning: Early childhood exemplars: Book 18 mathematics pāngarau*. Learning Media.

Ministry of Education. (2011). *Tātaiako: Cultural competencies for teachers of Māori learners*. Ministry of Education.

Ministry of Education. (2017a). *Te Marautanga o Aotearoa*. Ministry of Education.

Ministry of Education. (2017b). *Te whāriki. He whāriki mātauranga mō ngā mokopuna o Aotearoa: Early childhood curriculum*. Ministry of Education.

Ministry of Education. (2018a). *2017 Ngā Kura o Aotearoa New Zealand schools*. https://www.educationcounts.govt.nz/__data/assets/pdf_file/0005/188078/NZ-Schools-2017.pdf

Ministry of Education. (2018b). *Our purpose and vision*. https://www.education.govt.nz/our-work/our-role-and-our-people/our-purpose-and-vision/

Ministry of Education. (2019a). *e-ako*. https://e-ako.nzmaths.co.nz/

Ministry of Education. (2019b). *Initial teacher education statistics*. https://www.educationcounts.govt.nz/statistics/tertiary-education/initial-teacher-education-statistics

Ministry of Education. (2019c). *Tapasā: Cultural competencies framework for teachers of Pacific learners*. Ministry of Education.

Ministry of Education. (2019d). *Treaty of Waitangi*. http://nzcurriculum.tki.org.nz/Principles/Treaty-of-Waitangi

Murphy, C. (2012). The role of subject knowledge in primary prospective teachers' approaches to teaching the topic of area. *Journal of Mathematics Teacher Education, 15*(3), 187–206. DOI: 10.1007/s10857-011-9194-8

Murphy, C. (2016). Changing the way to teach maths: Preservice primary teachers' reflections on using exploratory talk in teaching mathematics. *Mathematics Teacher Education and Development, 18*(2), 29–47.

Nolan, K., & Walshaw, M. (2012). Playing the game: A Bourdieuian perspective of preservice inquiry teaching. *Teaching Education, 23*(4), 345–363.

Parra, A., & Trinick, T. (2018). Multilingualism in indigenous mathematics education: An epistemic matter. *Mathematics Education Research Journal, 30,* 233–253.

Peterson, E. R., Rubie-Davies, C., Osborne, D., & Sibley, C. (2016). Teachers' explicit expectations and implicit prejudiced attitudes to educational achievement: Relations with student achievement and the ethnic achievement gap. *Learning and Instruction, 42,* 123–140.

Pfannkuch, M., & Ben-Zvi, D. (2011). Developing teachers' statistical thinking. In C. Batanero, G. Burrill, & C. Reading (Eds.), *Teaching statistics in school mathematics-challenges for teaching and teacher education: ICMI Study Series* (pp. 323–333). Springer.

Pomeroy, D. (2018). Educational equity policy as human taxonomy: Who do we compare and why does it matter? *Critical Studies in Education.* DOI: 10.1080/17508487.2018.1440615

Rawlins, P., Anthony, G., Averill, R., & Drake, M. (2019). Novice teachers' perceptions of the use of rehearsals to support their learning of ambitious mathematics teaching. *Asia - Pacific Journal of Teacher Education, 48*(4), 389–402. DOI: 10.1080/1359866X.2019.1644612

Scoop Media. (2007). *Te Marautanga o Aotearoa Maori curriculum launch*. http://www.scoop.co.nz/stories/PA0711/S00292/te-marautanga-o-aotearoa-maori-curriculum-launch.htm

Shaughnessy, M., Ghousseini, H., Kazemi, E., Franke, M., Kelley-Petersen, M., & Hartmann, E. (2019). An investigation of supporting teacher learning in the context of a common decomposition for leading mathematics discussions. *Teaching and Teacher Education, 80*, 167–179.

Simon-Kumar, R. (2020). Justifying inequalities: Multiculturalism and stratified migration in Aotearoa/New Zealand. In R. Simon-Kumar, F. Collins, & W. Friesen (Eds.), *Intersections of inequality, migration and diversification* (pp. 43–64). Palgrave.

Sinnema, C., Meyer, F., & Aitken, G. (2017). Capturing the complex, situated, and active nature of teaching through inquiry-oriented standards for teaching. *Journal of Teacher Education, 68*(1), 9–27.

Stewart, G., Trinick, T., & Dale, H. (2018). Huarahi Māori: Two decades of indigenous teacher education at the University of Auckland. In P. Whitinui, C. Rodriguez de France, & O. McIvor (Eds.), *Promising practices in indigenous teacher education* (pp. 149–162). Springer.

Stockero, S. L., Rupnow, R. L., & Pascoe, A. E. (2017). Learning to notice important student mathematical thinking in complex classroom interactions. *Teaching and Teacher Education, 63*, 384–395.

Teaching Council of Aotearoa New Zealand [TCANZ]. (2019). *ITE programme approval, monitoring and review requirements*. Author.

Te Maro, P. (2018). *Māori engagement with pānga rau: Multiple mathematical relationships and te ao Māori*. (Unpublished doctoral dissertation). Te Whare Wānanga o Awanuiārangi, Whakatāne, New Zealand.

Thrupp, M., & White, M. (2013). *Research, analysis and insight into National Standards (RAINS) Project final report: National Standards and the damage done*. Wilf Malcom Institute of Educational Research.

Tomlins-Jahnke, H., & Warren, K. (2011). Full, exclusive and undisturbed possession: Māori education and the Treaty. In V. Tawhai & K. Gray-Sharp (Eds.), *Always speaking: The Treaty of Waitangi and public policy* (pp. 43–62). Huia.

Tomorrow's Schools Independent Taskforce [TSIT]. (2018). *Our schooling futures: Stronger together. Whiria ngā kura tūātinitini*. https://conversation.education.govt.nz/assets/TSR/Tomorrows-Schools-Review-Report-13Dec2018.pdf

Trinick, T. (2019, January–February). *Mathematics education: Its role in the revitalisation of indigenous languages and culture*. Paper presented at the 10th International Conference of Mathematics Education and Society, Hyderabad, India.

Turner, H., Rubie-Davies, C., & Webber, M. (2015). Teacher expectations, ethnicity and the achievement gap. *New Zealand Journal of Educational Studies, 50*(1), 55–69.

Tweed, B. (2015). *Tātai korero i ngaro, tātai korero i rongona: Legitimation and the learning of curriculum mathematics in an indigenous Māori school*. Unpublished doctoral dissertation. Victoria University of Wellington, New Zealand.

United Nations. (1989). *Convention on the rights of the child*. https://www.unhcr.org/uk/4aa76b319.pdf

United Nations. (2007a). *Convention on the rights of persons with disabilities*. https://www.un.org/development/desa/disabilities/convention-on-the-rights-of-persons-with-disabilities.html

United Nations. (2007b). *Declaration on the rights of indigenous peoples*. https://undocs.org/A/RES/61/295

University of Auckland. (2019). *Courses—Faculty of education and social work*: Education Curriculum Studies. https://www.calendar.auckland.ac.nz/en/courses/faculty-of-education-and-social-work/education-curriculum-studies.html

University of Otago. (2019). *EDUC477 secondary curriculum 1*. https://www.otago.ac.nz/courses/papers/index.html?papercode=EDUC477

University of Otago & NZCER. (2014). *National monitoring study of student achievement, mathematics and statistics 2013*. Ministry of Education.

Victoria University of Wellington. (2019). *TCHG 321: Mathematics and statistics education*. https://www.victoria.ac.nz/courses/tchg/321/2019

Ward, J., & Thomas, G. (2007). What do teachers know about fractions? In B. Annan, F. Ell, J. Fisher, N. Hāwera, J. Higgins, K. C. Irwin, A. Tagg, G. Thomas, T. Trinick, J. Ward, & J. Young-Loveridge (Eds.), *Findings from the New Zealand Numeracy Development Projects 2006* (pp. 128–138). Ministry of Education.

Whatman, J., & MacDonald, J. (2017). *High quality practica and the integration of theory and practice in initial teacher education*. New Zealand Council for Educational Research.

Wilson, S., McChesney, J., & Brown, L. (2017). Cultural competencies and planning for teaching mathematics: Preservice teachers responding to expectations, opportunities, and resources. *Journal of Urban Mathematics Education, 10*(1), 95–112.

Young-Loveridge, J., Bicknell, B., & Mills, J. (2012). The mathematical content knowledge and attitudes of New Zealand pre-service primary teachers. *Mathematics Teacher Education and Development, 14*(2), 28–49.

CHAPTER 7

EDUCATION OF TEACHERS WHO TEACH MATHEMATICS IN BRAZIL

Celi Espasandin Lopes
Universidade Cruzeiro do Sul

Adair Mendes Nacarato
Universidade São Francisco

In this chapter, we present and analyze the process of educating teachers who teach mathematics in Brazil and outline advances and challenges of mathematics teacher education. We present an overview of the organization of Brazilian education and the Brazilian scenario of teacher education for those who teach mathematics in Basic Education. We analyze the problems in public policies that impact education of teachers and discuss the pedagogy courses and the teaching degree in mathematics. Also, we highlight the diversity, complexity, and heterogeneity that mark the Brazilian context and interfere in the teacher education process.

INTRODUCTION

The purpose of this chapter is to present and analyze the process of educating teachers who teach mathematics in Brazil and outline advances and challenges of mathematics teacher education. We use the phrase "teachers who teach mathematics" to refer to all mathematics teachers, ranging from early childhood education teachers to university professors.

International Perspectives on Mathematics Teacher Education, pages 143–166.

Brazil is a country of continental dimensions; it has a territorial extension of 8,514,876 km², making it the fifth largest country in the world in terms of area. Its area corresponds to approximately 1.6% of the entire surface of the planet, occupying 5.6% of the dry land of the globe, 20.8% of the area of all the Americas, and 48% of South America. Its approximate population is 200 million people.

Lei de Diretrizes e Bases da Educação Nacional (National Education Guidelines and Framework Law) (Brazil, 1996), known as LDB, is the legislation that defines and regulates the organization of Brazilian education based on the principles in the Constitution. The first LDB was promulgated in 1961 (LDB 4024/61). In December 1996, a second bill was enacted—Law number 9.394/96, which has been in force until now and divides Brazilian education into two levels: basic education and higher education. This is the second time that education has had a Law of Guidelines and Bases for Education, which regulates all its levels.

In Brazil, Basic Education consists of three levels:

1. Early Childhood Education—day care centers (0–3 years) and preschools (4–5 years). These are free and compulsory for children from the age of 4 years, and are the responsibility of the municipalities.
2. Elementary School—initial years (from 1st to 5th grade, ages 6–10) and Middle School (from 6th to 9th grade, ages 11–14). These are compulsory and free, and are the responsibility of the municipalities and states.
3. High School—organized in three years (1st to 3rd year, ages 15–17). These are the responsibility of the states, and may include a vocational program.

Higher (tertiary) Education is the responsibility of the federal government and may be offered by those states and municipalities that have already met their responsibilities for levels of basic education. Both Basic and Higher Education can be offered by private institutions, which are overseen by state and federal governments.

Brazilian education also has some types of education that permeate all levels of national education:

- Special Education—serves students with special needs, preferably in the regular school system;
- Distance Education—serves students in different times and spaces, using information and communication technologies and media;
- Professional and Technological Education—aims to prepare students to perform productive activities, as well as to update and improve technological and scientific knowledge;
- Youth and Adult Education—serves people who have not had access to education at the appropriate age; and
- Indigenous Education—serves indigenous communities in a way that respects the culture and mother tongue of each community.

Despite all of this organization provided for in the legislation, which aims to meet all the demands of the Brazilian population, there are still numerous situations in which such services are not provided. Historically, Brazil has been characterized by great social inequality and cultural diversity. Ethnic and racial issues have been the object of many social struggles, such as the Black Movement in the 20th century. "Throughout its history, Brazil has set a model of exclusionary development, preventing millions of Brazilians from joining and remaining in school" (Brazil, 2004, p. 5). Such problems also affect teacher education courses throughout the national territory. These courses have contributed nothing to economic and cultural development and to overcoming serious social problems of Brazil.

In this chapter, we analyze in more detail the pressures and limitations of the preservice education of mathematics teachers. We will not cover all modes of education, as each would warrant a separate text. Our focus will be on the teaching degree in mathematics and the pedagogy major. This program choice is due to the fact that such courses inform the work of teachers in schools, public and private.

This chapter is organized as follows. First, we present an overview of the organization of Brazilian education and the Brazilian context of teacher education for those who teach mathematics in Basic Education. Second, we present and analyze a few public policies that contribute to the preservice education of teachers. Third, we discuss more broadly the two kinds of degrees: a pedagogy program that prepares teachers who teach mathematics in elementary school; and the teaching degree in mathematics that prepares teachers who teach mathematics in middle and high school. Finally, we analyze the problems in public policies that impact teacher education.

THE CONTEXT OF TEACHER EDUCATION IN BRAZIL

The initial education of mathematics teachers in Brazil is achieved through university courses overseen by the Ministry of Education. There are two types of teacher education programs: a mathematics teaching major, which prepares teachers to work in the second half of elementary school and high school; and a pedagogy program, which prepares teachers to work in early childhood education and in the first half of elementary school.

The latest LDB Law, number 9.394/96 (Brazil, 1996), instituted the requirement of university education for mathematics teachers. Following this legislation of a wider nature, other laws were enacted to set guidelines for teacher education. We focus on two of them:

1. Resolution CNE/CP of 2002 (Brazil, 2002a) implemented national curricular guidelines to train basic education teachers (early childhood to high school). As well as standardizing the structure of courses, this resolution focuses on research regarding the teaching and learning process and the introduction of teaching practice from the very beginning of the program, as well as a greater number of hours of supervised intern-

ship. However, this resolution did not produce advances for the courses that had already been implemented, as most preserved their pedagogical principles, which are the guiding practices and methodologies of the courses as well as the purposes, conceptions, and guidelines for their operationalization.

2.   Resolution no. 2, of July 1, 2015 (Brazil, 2015a), set forth guidelines for teacher education at the university level and for continuing education. These guidelines are an advancement in comparison to previous regulations. They propose integration between initial and continuing education and establish the workload of curricular components, such as Teaching Practice (400 hours), Supervised Internship (400 hours), and theoretical and practical activities (200 hours). The guidelines set a minimum workload of 3,200 hours for the program, distributed throughout eight semesters, both for obtaining a degree in mathematics and for obtaining a degree in pedagogy.

The Brazilian higher education system is comprised of public and private institutions. Public institutions are the universities (owned and administered by federal, state, or municipal governments) and education institutes. The education institutes were created in 2008 to provide vocational education to high school students (in the 15–17 age range). These institutes offer vocational courses concomitantly with high school, so students who complete such courses can continue to higher education. The institutes also offer college courses, and 20% of these institutes must provide teaching degrees, including pedagogy. They also offer post-graduation courses (specialization courses, master's, and doctorate degrees).

In the private sector, there are three types of institutions: universities, university centers, and colleges. Universities have academic autonomy and are based on the teaching/research/continuous professional development tripod. University centers have no academic autonomy, but they can conduct research if they offer post-graduation courses. Colleges only offer undergraduate studies.

DIVERSITY IN THE BRAZILIAN CONTEXT

The beginning of the 21st century was characterized by some advances in education. In 2004, the National Education Council (CNE) approved the National Educational Guidelines for Ethnic/Racial Relations and for Teaching Afro-Brazilian and African History and Culture. These guidelines specify that "valuing race identity and ethnicity is one of the relevant arguments that justify the need to contextualize them within school curricula, which should encompass the entire Brazilian population" (Almeida et al., 2018, p. 6). Therefore, the recommendation was to prepare teachers in ways that ensure the inclusion of all students in elementary education. Before these guidelines were enacted, the federal government created the National Affirmative Action Program through Decree no. 4.228, on May 13, 2002 (Brazil, 2002b). The decree ensures the participation of African descendants, women, and disabled individuals in commissioned positions of Management and Consulting (DAS). Enforcement of this decree is delegated to the Secretary of Human Rights of the Ministry of Justice. Later, on August 29, 2012, Law no. 12.711 (Brazil, 2012) instituted the quota regimen. Through that regimen, self-declared black, mixed-race, and indigenous individuals, as well as individuals with disabilities, had guaranteed access to a percentage of the places at universities. These actions have ensured enrollment at universities of a significant number of working-class students, mostly African descendants and students graduating from public schools. Before quota laws, the public university system was comprised mainly of middle-class white students.

In the case of indigenous populations, since 2008, federal and state universities have created and provided teaching degrees in indigenous studies. This program, known as the Higher Education and Indigenous Degree Program (*Programa de Formação Superior e Licenciaturas Indígenas* - Prolind) and created from the National Education Plan (Plano Nacional de Educação) Brazil (2001), was enacted on January 9, 2001. It features a chapter on indigenous school education, divided into three parts. The first part provides a quick analysis of how school education has been offered to indigenous peoples. The second part has guidelines for how indigenous school education is presented. The third part lists short and long-term objectives and goals that must be achieved to support the university education of teachers who work in elementary education at indigenous schools. The purpose of this program is to prepare these educators to teach in the final years of elementary school (12–15 years of age) and high school. The pedagogical principles of these courses, as well as being different from place to place, respect the intercultural and territorial differences of each ethnic group. The program lasts four years; during that period, indigenous students alternate their time between the university and their communities. The pedagogical principles promote dialogue between traditional indigenous knowledge and academic knowledge, in order to build meaningful pedagogical work.

Also, in order to solve educational inequalities, in 2002 the Ministry of Education established the Support Program for Teacher Education in Rural Education (*Programa de Apoio à Formação Superior em Licenciatura em Educação do Campo* - Procampo) (Brazil, 2010). The mission of Procampo is to promote education for teachers who work in the public network of rural schools, as well as educators who participate in alternative experiences, that is, programs that differ from what is determined for regular education. These programs better serve the learning needs of students in a particular community, in this case rural communities, so as to expand the opportunities for quality elementary education for the students who live in these areas.

Thus, there is a range of programs offered for preservice education for teachers: a pedagogy program, which certifies multi-subject teachers to work in early childhood education and the initial years of elementary school; mathematics teaching degrees, which afford single subject teaching credentials; indigenous intercultural teaching degrees; and teaching degrees for rural education. These different manners of preparation are the result of struggles within society to reduce educational inequalities and, as a consequence, social inequalities, as well as value the multiplicity of cultures.

However, particularly after the impeachment of President Dilma Rousseff, on August 31, 2016, educational policies underwent setbacks, due to new practices and actions implemented by the federal government. The new educational programs and guidelines are now defined in a more centralized manner, without broad participation of organized civil movements, as well as scientific societies and academic representation. Significant setbacks in the agenda and in public educational policies impacted topics related to research funding and educational programs, evaluation, management, curriculum, and teacher education. As well, there was a conservative and privatizing counter-reform through a vast process of deregulation, which has favored the expansion of private enterprises, disregarding resources produced by educational agencies and institutions committed to the quality of public education (Dourado & Oliveira, 2018).

So far, this legislation has not yielded significant changes in the conception of preservice education, particularly of mathematics teaching degrees. The debates have been intensive, and the advancements have not been very significant. In the case of teachers who specialize in mathematics and who teach in the final years of elementary school and high school, education traditions are hard to break. Such degrees are still plagued by the notion of a teaching degree focused on pedagogy as a second-class bachelor's degree; i.e., the curriculum has a strong presence of the discipline of pure mathematics, with a reduced workload geared towards Education Sciences, which would ensure the pedagogical background of future teachers. In the case of pedagogues, multi-subject teachers who work in early childhood education and the initial years of elementary school, preservice education is centered on Education Sciences, with little emphasis on teaching methodologies for mathematics.

PREPARATION OF TEACHERS WHO TEACH MATHEMATICS IN ELEMENTARY SCHOOL[1]

In this section, we address university teacher education. Initially, we focus on the pedagogy degree program, then, on the teaching degree in mathematics. Both types of teacher education are offered on-site or through distance learning.

Since the LDB bill enacted in 1996, teacher education for early childhood education and the early years of elementary school—the so-called multiple subject teacher—has taken place through a university degree in pedagogy. The program is comprised of 3,200 hours of general instruction. Although the central focus of this education is geared towards teaching, the degree also enables graduates to work in the area of school management or other areas, such as business pedagogy

---

[1] Recall that in Brazil, elementary education encompasses grades 1 to 9.

(for professionals who will work in companies), and hospital pedagogy (for those working in hospitals).

Taking into consideration national diversity and the pedagogical autonomy of the institution, each institution develops its pedagogical principles at its own discretion, comprised of the following core:

1. *General preparation studies*, encompassing specific and interdisciplinary areas, their foundation and methodologies, and various educational realities;

2. *Deepening and diversification*, studies regarding professional activity, including specific and pedagogical content, prioritized in accordance to the pedagogical principles of each institution, though in tune with the educational system; and

3. *Integrative studies for curriculum enrichment.*

Thus, over the course of four years, preservice teachers are exposed to Education Sciences in general, such as Philosophy of Education, Sociology of Education, Developmental Psychology, and General Didactics, among others; and specific methodologies related to basic education, such as Portuguese and literacy, history and geography, science, mathematics, art, and physical education. Additionally, there are courses focused on the education of adolescents and adults, Brazilian Sign Language (*Libras*), Inclusive Education, and other specific content. An example of the curriculum of a pedagogy program at the Federal University Rio Grande do Sul (UFRS) in the southern region of Brazil is shown in Figure 7.1. The curricular guidelines aim to contemplate the totality of the education of the pedagogue and is articulated around six axes: scientific-intellectual formation; fundamentals of education; learning and teaching processes; educational management; education and diversity; and pedagogical practice. Despite the classification of curricular teaching activities on the different axes, several of them promote integration between them. Another important element is that each axis is not focused on a specific moment in the course but is distributed along the student's trajectory in the search for greater integration of the training process. For this program, the specific case of teaching methodologies for mathematics, a single two-semester long course is usually offered, which has a workload of about 60 hours.

A large research project carried out by Gatti and Barretto (2009), with a grant from UNESCO, analyzed the course lists and pedagogical projects of several pedagogy courses, and found the following:

> Among public universities, none of those surveyed offered courses on substantive content of each area of knowledge, not even Portuguese and Mathematics. Such content remains implicit in the courses pertaining to teaching methodologies, or in the presumption that teachers-in-training already have a good grasp of the syllabus. We were able to identify the treatment of specific content to be taught in elementary schools in just a few course lists. (p. 126)

Course Title	Number of Hours
Semester 1	
Education and Society	60
Musical Education	45
History of Education I: Modernity and Interculturality	60
Reading and Written Production in Teacher Constitution	45
Policy and Organization of Basic Education	60
Psychology of Education: Constitution of the Subject	45
Semester 2	
Education and Cinema	45
Special Education and Inclusion	45
Education Philosophy I: Foundations of Pedagogical Experience	60
History of Education II: Subjects, Institutions and Practices	45
Games and Education	45
Brazilian Sign Language (Libras)	30
Education Psychology: Knowledge and Learning	60
Semester 3	
Acquisition and Development of Oral and Written Language	60
Youth and Adult Education: Practices and Subjects	45
Education and Visual Arts	45
Social Education: Fundamentals and Practices	45
School Management	60
Education Sociology: Theoretical Background	60
Semester 4	
Literacy: Theoretical-Methodological Concepts	60
Education and Theater	45
Special Education, Education and Inclusive Processes I	45
Child Education: Practices and Subjects	45
Media, Digital Technologies and Education: Learning Processes and Methods	45
Semester 5	
Contemporary Education: Curriculum, Teaching, Planning	60
Mathematical Education I	75
Teaching Internship Seminar: Special Education, Teaching and Inclusive Processes in Special Education	30
Teaching Internship Seminar: Special Education, Processes and Practices	30
Teaching Internship Seminar: School Management	30
Teaching Internship: Pedagogical Practices in Social Education and Rural Education	30
Semester 6	
Early Childhood Education: Practices and Subjects	45
Education in Natural Sciences	60
Mathematical Education II	45
Education, Health and Body	30
Psychopedagogy	45
Sociology Education: The School in Focus	45
Semester 7	
Education and Ethnic/Racial Relations	30
Teaching History	45
Gender and Sexuality in Education	30
Space Reading: Geography for Early Childhood	45
Research in Education	45
Senior Capstone I	30

Course Title	Number of Hours
Semester 8	
Teaching Internship Seminar I: Elementary School—initial years	75
Teaching Internship Seminar I: Youth and Adult Education	75
Teaching Internship Seminar I: Early Childhood Education	75
Teaching Internship Seminar II: Elementary School—Initial Years	75
Teaching Internship Seminar II: Youth and Adult Education	75
Teaching Internship Seminar II: Early Childhood Education	75
Semester 9	
Philosophy of Education II: Philosophical Problems	60
Senior Capstone II	60

FIGURE 7.1. Sample Curriculum of Pedagogy Program of Universidade Federal do Rio Grande do Sul (UFRGS) Source: UFRGS (2018)

It can be said that the majority of courses directed to teaching mathematics focus on methodological aspects and not on epistemological and conceptual foundations. In addition, in their analysis, the authors found that the mathematical content offered in such courses is focused on numbers, the four fundamental operations, fractions, and problem solving.

This mirrors the conceptual fragility displayed by pedagogy graduates, as pointed out by the authors: "The insufficient preparation of preservice teachers regarding traditional school content, even at basic level, requires a more in-depth reflection on the sufficiency or adequacy of the education of multi-purpose teachers, as well as the interdisciplinary perspective" (Gatti & Barretto, 2009, p. 128). In addition to this insufficient preservice preparation, the authors outline the profile of teacher trainers in pedagogy courses. In most institutions, mathematics educators are not in charge of the course Methodology for Teaching Mathematics, but rather pedagogues who, by way of their own education, display conceptual gaps in mathematical knowledge. Few institutions have mathematics educators in the pedagogy program.

Most pedagogy courses are offered as evening classes, and according to the latest Higher Education Census, conducted in 2018 (INEP, 2019), around 62% of these courses are concentrated in private educational institutions. In such institutions, lecturers are hired solely for teaching, and most have no link with research.

In addition to this scenario, in recent years, there has been enormous growth in long-distance teacher education. The highest percentage of enrollments is in pedagogy; according to the latest census in 2018, 26% of total enrollment was for this major. Most of these courses are offered by private institutions. According to Gatti et al. (2019, p. 126), "In the distance education mode, for the pedagogy major, every 23 enrollments in private institutions correspond to less than two in public institutions." The pedagogy program educates the generalist teacher who can teach any content in early childhood education or primary school.

Regardless of the type of program, on-site or distance education, it has been well established that pedagogy program graduates do not have the repertoire of mathematical knowledge needed to work in the levels of education for which they have been trained. The gap is significant; educational companies, aware of this scenario, have been investing in the production of teaching materials for use in the classroom under a perspective that is more concentrated on techniques and skills than reflection. These companies produce *how-to* manuals as a way to mitigate these conceptual gaps in teacher education, disregarding all the discussion in research about the need for more meaningful preparation and ignoring significant production in the field. In Brazil, the largest research production in education concerns teacher education. Researchers in different areas of knowledge have turned their attention to the issue of teachers, relying on extensive national and international literature. As Nóvoa states:

> We are faced with a kind of *discursive consensus*, rather redundant and long-winded, which is multiplied in references to the professional development of teachers, the articulation of pre-service education, induction and in-service education, in a lifelong learning perspective, to the attention to the early years of practice and the inclusion of young teachers in schools, the idea of reflexive teachers and teacher education based on research, the new skills of 21st century teachers, the importance of collaborative cultures, teamwork, supervision and evaluation of teachers ... and so on. (2007, pp. 2–3)

Given the difficulties in the dialogue between the academic community and public policy makers, the void has been occupied by educational entrepreneurs, in a neoliberal conception, when education leaves part of the social and political field to enter the market. Neoliberal rhetoric links school education to job preparation and market imperatives; the school becomes a means of transmitting doctrinal principles and becomes a market for the products of cultural and computer industries. This perspective also makes teachers responsible for their professional development, thereby releasing the State of the responsibility for teacher education and delegating it to other sectors. Thus, teacher education is relegated to business sectors and a "responsibility policy," through which teachers become responsible for the low performance of their students in external evaluations. This happens because this neoliberal model of education operates within concepts of effectiveness, skills, and meritocracy, in flagrant control of teaching practices. According to Freitas (2014):

> Regarding educators, the same disqualification occurs within the productive processes, with the introduction of new technologies and managerial controls, that is, professionals are becoming easily replaceable, thus their hasty training. (p. 1102)

In the context presented here, regarding the profile of the teacher who works in early childhood education and the first grades of elementary school, we agree with Gatti et al. (2019):

The hierarchy among those educated in elite institutions who occupy the center of the system and qualify for the most prestigious and well-paid jobs and the institutions dedicated to absorbing enrollments of low-income youths who get low-paying jobs is maintained. Among such degrees are, above all, pedagogy courses, which accredit for working with young children those most relegated to second-class diplomas. (p. 137)

If the challenges for the program in pedagogy are immense, even more complex are those of the major in Mathematics. Similarly, these programs are offered by public (universities and federal institutes) and private institutions (universities, university centers, and colleges) both on-site and through distance education.

PREPARATION OF TEACHERS WHO TEACH MATHEMATICS IN MIDDLE AND HIGH SCHOOL

The teaching degree in mathematics follows the national curricular guidelines for training basic education teachers, which requires a workload of 3,200 hours, distributed over eight semesters. Each institution has the autonomy to distribute the different courses during this period, provided that some legal requirements are met, especially regarding hours of practice and internship. There is a disparity in the curriculum configuration of the different institutions and official data are virtually non-existent. An example of the curricular grid of the Federal University of Juiz de Fora, which is located in the southeastern region of Brazil in Minas Gerais, is shown in Figure 7.2.

In Brazil, the education of mathematics teachers is linked to two scientific societies, which do not share the same teaching and learning perspectives, thereby generating tensions. On the one side is the Brazilian Mathematical Society (*Sociedade Brasileira de Matemática*—SBM), which is traditionally linked to the baccalaureate in mathematics, and in the departments of universities represented by professors who teach courses focused on pure mathematics. On the other side is the Brazilian Society of Mathematics Education (*Sociedade Brasileira de Educação Matemática*—SBEM), which represents mathematics educators.

Research by Gatti and Barretto (2009) revealed that among the course programs for prospective teachers that were analyzed, practically one third (32.1%) of the workload was devoted to specific content (pure mathematics), 12.8% for Education Sciences, 30% for teacher education, and 25.1% for other components (senior capstone, research, or complementary activities). The authors found that mathematics courses have not yet incorporated in their curricular matrices a significant workload contemplating important aspects of the training of teachers who will work in early childhood education and elementary school. As an example, one gap in the syllabus of undergraduate courses in mathematics is the lack of discussion about external evaluations (SAEB, SARESP, ENEM, PISA) as well as the low scores achieved by students on such evaluations. Evaluating students is not a trivial matter for educators, but requires training and discussion. However,

Course Title	Number of Hours
Semester 1(360 hours)	
Calculus I	60
Analytical Geometry and Linear Systems	60
Fundamental Chemistry	60
Algorithms	60
Programming Laboratory	30
Laboratory of Introduction to Physical Sciences	30
Chemistry Laboratory	30
Introduction to the Exact Sciences	30
Semester 2 (300 hours)	
Calculus II	60
Physics I	60
Physics Laboratory I	30
Structure and Transformations Laboratory	30
Introduction to Statistics	60
Fundamentals of Elementary Mathematics	60
Semester 3 (330 hours)	
Calculus III	60
Physics II	60
Introduction to Number Theory	60
School Mathematical Knowledge	60
School Practice in School Mathematical Knowledge (corresponds to teaching methodologies and practices in mathematics)	30
State, Society and Education	60
Semester 4 (390 hours)	
Differential Equations I	60
Linear Algebra	60
Plane Geometry	60
Informatics in Mathematics Teaching (use of software that aids mathematical learning)	60
Mathematics Teaching Methodology	60
Public Policy and School Space Management	60
School Ethics in Public Policy and Management School Space	30
Semester 5 (330 hours)	
Introduction to Mathematical Analysis	60
Trigonometry	60
Spatial Geometry	60
Process of Teaching and Learning	60
Mathematics Teaching I (only methodologies, no teacher practices)	30
School Practice in Mathematics Teaching I (corresponds to teaching methodologies and practices for mathematics students in the age group of 11 to 14 years)	60
Probability Calculation	60
Semester 6 (390 hours)	
Complex Functions	60
Exponentials and Logarithms	60
Financial Mathematics	60
Geometry Topics	60
Mathematics Teaching in Basic Education II	30
School Practice in Mathematics Teaching II (corresponds to teaching methodologies and practices in mathematics for students in the age group of 15 to 17 years)	60
School Mathematics I (elementary mathematics for students in the age group of 11 to 14 years)	60

Course Title	Number of Hours
Semester 7 (500 hours)	
Algebraic Structures	60
Discrete Mathematics	60
School Mathematics II (elementary mathematics for students in the age group of 15 to 17 years)	60
Elective Discipline	60
Philosophical Issues Applied to Education	60
Supervised Internship in Mathematics Teaching I	140
Reflections on Performance in School Space I	60
Semester 8 (440 hours)	
School Mathematics II (elementary mathematics for students in the age group of 15 to 15 to 17 years)	60
Mathematics History	60
LIBRAS and Education for the Deaf	60
Elective Discipline	60
Supervised Internship in the Teaching of Mathematics II	140
Reflections on Performance in School Space II	60
Curricular Flexibility Activities (200 hours, developed throughout the program)	200

FIGURE 7.2. Sample Curriculum for the Mathematics Teaching Program Source: https://www.ufjf.br/matematica/files/2015/04/PPC.licenciatura.novembro.pdf (pp. 21–24). Accessed October 13, 2019

as far as could be verified, graduates with mathematics majors, as well as graduate students in other fields, do not receive education for that.

In most institutions, teaching certification courses are still based on a classic teacher education model, which has prevailed in Brazil since the 1930s, known as 3 + 1. The student has 3 years of specific education in mathematics (towards a bachelor's degree) and 1 year of didactic-pedagogical education. There have been attempts to break with this model of teacher preparation, as well as discussions about which kind of mathematics should be taught in the undergraduate course. Moreira and Ferreira (2013), in a public debate on such issues, argue the following:

> The national and international consolidation of mathematics education as a field of study and the consequent development of research literature specialized in the education of mathematics teachers have decisively broadened the understanding regarding the knowledge about teaching practice and, to the same extent, knowledge which is potentially relevant to pre-service teacher education. (p. 984)

Among such knowledge, the importance of specific content cannot be denied, but content alone is not enough. The social contexts of basic education are increasingly complex, with numerous classes of large size (about 40 students) and with marked heterogeneity and wide cultural and social diversity. Schools face many new demands and teachers need to have a more encompassing and critical education in order to deal with such issues.

Moreira and Ferreira (2013) point to two different trends for answering the question of which kind of mathematics should be included in teacher education. The first is studies that "seek to encompass mathematical knowledge relevant to the professional practice, in terms of the specificities of teaching mathematics, and not preponderantly the academic discipline itself" (p. 999). Such studies reference Shulman (1986), in connection with the PCK (Pedagogical Content Knowledge) category. Among those are the works of Ball et al. (2008) in the USA, and Moreira and David (2005) and Moreira and Ferreira (2013) in Brazil. We could also mention the influences of the studies of Carrillo et al. (2013) in Spain. The other trend is derived from a view that is closer to the 3 + 1 model of teacher education, which emphatically values knowledge of content, viewing it as the ultimate core of teacher education. Moreira (2012) states that specific content courses continue to be taught independently of other course subjects, which does not guarantee the professional knowledge required for teaching:

> Teacher training courses have moved away from the 3 + 1 formula. However, how can we state that 3 + 1 did not move away from teacher education? The answer is that the underlying logic of 3 + 1 is still present as the logic which permeates the structure of such courses. The fundamental principle is still the same: the separation between content courses and teaching courses. What clearly changed was the composition of the group of subjects related to education (which in 3 + 1 was practically solely didactics) and the proportion between the training periods encompassed by the scientific content and the teaching/education group. Now, this ratio revolves around 1:1. Once the 3 + 1 underlying logic has been internalized and naturalized, this new ratio may seem like a pretty radical change, however, in my view, the crucial issue remains unstirred. (p. 1140)

This conflict about what kind of mathematics should be offered during preservice teacher education is also reinforced by the profile of the educators who work in undergraduate courses. These mathematics teacher educators, for the most part, are connected either to SBM or SBEM, which reinforces their perspectives about what mathematics is, and how the teaching and learning of mathematics should occur. Thus, there are clashes within and outside the courses. Relationships between the two societies are often in conflict, and collaborative work is not achieved. Members of SBM believe that teacher education resources should be invested in teaching mathematical content. In contrast, members of SBEM believe that this is not enough; there is a need for a more encompassing education, focusing on pedagogical knowledge of the mathematical content, as well as on the cognitive characteristics and particularities of each student age range.

Diniz-Pereira (2011) lists characteristics of teacher education models, as summarized in Figure 7.3. Such models coexist, as they are not mutually exclusive in the concreteness of teacher education and practice. It so happens that the perspective of SBM is more closely related to the technical rationality model, whereas SBEM bears closer proximity to the models of practical and critical rationality. In this way, the community of mathematics educators maintains a constant discussion about preservice courses. Every two years, SBEM organizes the national forum on education of teachers who teach mathematics, which is preceded by regional forums. During such meetings, the problems and advances in education offered by pedagogy and mathematics courses are discussed, and successful experiences are shared. However, the community recognizes that these discussions have no echo in the creation of public policies, as SBEM is rarely called upon to debate ideas.

This is, no doubt, a paradox, as the amount of research is extensive and most of it receives funding from public agencies. With regard to research developed in Master's and doctoral programs, a national mapping identified 858 research projects conducted in the period from 2002 to 2012 (Fiorentini et al., 2016). However, what is produced as a result of that research has little influence on policies.

Currently, SBEM, through the working group on teacher education (WG07), is conducting nationwide research to diagnose which mathematics is being taught in the programs of pedagogy and teaching degrees in mathematics, both on-site and through distance education. So far, research had been carried out by individuals or research groups, who addressed more specific issues regarding the mathematical education of such professionals.

Teaching degree programs in mathematics are also offered through distance education, mostly in private institutions. Gatti (2014) draws attention to both the

Teacher Education Model	Characteristics	Conception of Teacher
Technical Rationality	Emphasizes training of behavioral abilities and is structured according to the transmission of the so-called scientific content, viewed as sufficient for teaching.	The teacher is seen as a technician who, in an objective way, must put into practice the scientific and pedagogical knowledge that they have acquired during education.
Practical Rationality	Practice cannot be reduced to a sum of predictable, theory-controlled events. Knowledge is not reduced to reproductive control of student management in order to obtain the expected results. Broader aspects of the school context, such as social and moral relations, are valued, but are still far from transformative propose.	The teacher is encouraged to reflect on the problems of practice and guide their actions according to pedagogical justifications. The teacher is a reflective professional.
Critical Rationality	Praxis plays a crucial role in the unveiling of reality.	A problematizing perspective is encouraged, through which the teacher constructs knowledge with the students, based on their needs, seeking the transformation of reality and promotion of social justice.

FIGURE 7.3.   Teacher Education Models

quality of such programs and the care required with students, especially considering the profile of young people seeking teaching degrees.

> Also, students in distance education do not benefit from contact with academic cul-
> ture, such as direct dialogue with colleagues in their own and other areas of knowl-
> edge, daily with teachers, through participation in student movements, debates, and
> a variety of experiences that university life offers more intensively. That is, these
> teachers in training lack a cultural socialization which is not negligible. Evidently,
> such characteristics are associated, for the most part, with the way distance educa-
> tion projects for teacher education are presented in the country. (p. 37)

As mathematics educators, we defend a view of education that encompasses
teacher education based on contributions of reflection and experiences. We consider these to be essential principles for the educational process, which demands
a context where problematization of reality and listening are the foundations of a
dialogue that results in practice. In the case of processes of teachers' education,
learning is twofold and simultaneous. The learning occurs among teacher educators and student teachers, at universities, and within the teacher-student relationship, in elementary and secondary education, at all stages and modes, encompassing all age groups: children, youth, and adults. Thus, learning and teaching to
listen is content for both teacher education and for professional practice (Santiago
& Batista Neto, 2011).

> Listening is obviously something that goes beyond one's auditory abilities. Listen-
> ing, in the sense discussed here, entails the permanent willingness on the part of
> those who listen, and openness to heed the speech, the gestures, the differences of
> others. This does not mean, of course, that listening demands that those who actually
> listen yield to those who speak. (Freire, 1991 as cited in Santiago & Batista Neto,
> 2011, p. 11, p. 135)

Listening as both an attitude and syllabus for education and professional practice
is evident in the work of Freire (1996), as can be observed when he refers to those
who seek education. In teacher education, listening is within the scope of the necessary knowledge for educational practice, not only learned and apprehended, but
exercised in the educational process of teachers. Listening is both an attitude and
content of educational practice that promotes a relation of respect between teacher
and student when exercising the right to speak and listen, which characterizes the
dialogical practice (Freire & Shor, 1986). In the next section we discuss some
public policies that could have positive effects on teacher education.

PUBLIC POLICIES FOR TEACHER EDUCATION

In Brazil, the Ministry of Education articulates and develops public policies that
must be approved by the legislature in order to come into force. Here, we highlight two federal programs: the Scholarship Mentorship Initiative Program (*Pro-*

grama Institucional de Bolsas de Iniciação à Docência - Pibid) (Brazil, 2016b) and Pedagogical Practicum (*Residência Pedagógica*) (Brazil, 2018).

Pibid presents an interesting educational alternative experience for the student teacher to facilitate experiencing the reality of mathematics classrooms in schools. Through a partnership between universities and schools that promotes collaborative work, students in education work in elementary, middle, and high schools under the supervision of a licensed teacher of the school who participates in the continuous education programs of the university. Both the undergraduates and teachers receive a scholarship.

Pibid involves undergraduates who are pursuing teaching majors in actions, practice, and educational activities, while experiencing the daily routines of schools and the challenges that emerge in the educational space. Regarding a mentorship program for introducing teachers in training to schools, Silva and Cruz (2018) consider that Pibid enables teachers in training to build a professional identity based on dynamics through which socialization and interactions occur seamlessly. In this context, the learning experiences of the teacher mediate the whole process of initial education, which presents itself as an historical and social construction, capable of producing sense and meaning to teaching, thus fostering the professionalization process (Garcia, 1999).

This is an effort of public authorities that yields very positive results as indicated by the amount of research produced in the last decade. However, Pibid has changed, having been mischaracterized in recent years in relation to its initial goals. Currently, there are many educators involved, but there is some apprehension about the path it is taking. It is feared that it will be even more mischaracterized, given the lack of teachers in public schools; this may require the participants of the program to teach classes to make up this gap, thereby causing it to fail to meet its objective of establishing teacher-training partnerships.

The Pedagogical Practicum Program was implemented in 2019 with an aim to foster the improvement of practical training in undergraduate courses by promoting the immersion of undergraduates working towards teaching licenses in elementary education during the second half of the course sequence required for a teaching major. It is an institutional program involving professors, students, and teachers and was created to improve the practicum with a scholarship and an extra semester for initial education, with a minimum duration of 1,600 hours. Practicum grants can be offered to graduates up to three years after completion of undergraduate teaching courses. Each professor, a teacher educator, is responsible for a group of students and teachers from the elementary, middle, and high school.

The programs, Pibid and Pedagogical Practicum, need to be related to other educational policies, such as, for example, the production of curricular documents, in order to generate effective improvements in the teaching and learning process. Successful experiences in these programs within teacher education should be a reference for public policy. However, unfortunately, it is not being considered by the Brazilian Ministry of Education.

Since 2015, the Brazilian educational community has been apprehensive due to the enactment of the National Common Curricular Core (*Base Nacional Comum Curricular - BNCC*). BNCC has three versions, with the final being published in 2017 (Brazil, 2015b, 2016a, 2017). Under a conception of neoliberal policies, BNCC determines the skills and abilities to be developed in each discipline of the curriculum and in each school grade. The document, whose implementation will be completed in 2020, has met much criticism from the community. In addition to standardizing the work of teachers and the production of teaching materials, it remains bound to large-scale external evaluations, which gauge students' performance and evaluate the work of the teacher, in a blatant attempt to control pedagogical work. The publication of this curriculum document greatly affects teacher education courses as:

> Many things are set in motion. The continuing teacher education and initial training courses for teachers begin to discuss the documents, textbooks are redesigned, research studies in the area are being developed, partnership projects between universities and schools are encouraged, and teachers contemplate the changes offered by the new curriculum. These efforts require great financial investments by municipal, state, and federal governments. (Lopes & Grando, 2018, p. 204)

However, to date, there have been no governmental actions to promote discussions and possible curricular adaptations that benefit the education of children and young people. Public policies have prioritized demands that favor private K–12 educational institutions. There is a movement in support of marketing actions in education, which are close to a utilitarian view of education, basically geared towards corporate profit.

As a result, opposition to the curriculum prescribed by the Ministry of Education rendered the tenability of these curricular documents and public policies impossible. This is aggravated by the tradition in Brazil that every change of government in states and municipalities is followed by changes in both the curriculum and educational management (Lopes & Grando, 2018).

BNCC reinforces the logic of competency-based learning, with restrictive curricular design strongly articulated with standardized assessment (Dourado & Oliveira, 2018).

> The political-pedagogical vision, ingrained in the structure of BNCC, does not guarantee or ratify national identity, under the notion of pluralism of ideas and pedagogical conceptions, valuing and respecting diversity and effective inclusion, knowledge, cultural and artistic values, at national or regional levels. These propositions significantly alter the regulatory framework for Elementary Education and in Higher Education, and are directly related to the organization, management and regulation of teacher education. (p. 41)

Due to these curricular perspectives and the discontinuities of effective public policy programs, we see the country aligned with the guidelines of international mechanisms such as the World Bank and the Organization for Economic Cooperation and Development (OECD). A visible example of such orientation lies in the views towards mathematical literacy. Brazil has a large repertoire of studies on reading and writing skills and literacy, with emphasis on the works of Paulo Freire and other contemporary studies, which value the mathematical knowledge produced through multiple social practices, considering the cultural diversity of the country. However, the BNCC guidelines dismiss the national production and adopt the Mathematical Literacy Matrix concept of PISA 2012, with adaptations. Passos and Nacarato (2018), when comparing BNCC and PISA, found that mathematical literacy is the individual's ability to formulate, employ and interpret mathematics in a variety of contexts, which refers to a definition centered on skills and abilities, in line with the neoliberal views of education prevailing in the country.

In view of the elaboration of the Common National Core for Teacher Education, the implementation of BNCC becomes not only a curricular issue, but directly affects the initial education of teachers in regular programs—with the exception of rural and indigenous education programs, which manage their own training and whose education must be in line with the skills and abilities to be developed with elementary, middle, and high school students. Undoubtedly, this constitutes a hegemonic and standardized model of public education, which clearly disregards local and regional possibilities and contexts.

FINAL CONSIDERATIONS

We believe that the ideas presented here outline a panorama of the courses for majors in pedagogy and mathematics in Brazil. In addition to the above-mentioned challenges, there is also the issue of the interest of young people in a teaching career. A teaching career has become unattractive to young people today and the profile of students seeking teacher education programs has changed over the years. Currently, most of these students come from regular high school courses in public schools; their access to public higher education has been guaranteed by an Affirmative Action Quota Regimen, which since 2012 guarantees 50% of enrollment places for this population. In private institutions, access has been made possible by the "University for All" program (*Programa Universidade para Todos* - Prouni), created in 2005. This program finances, with scholarships of up to 100%, the higher education of low-income university students who have graduated from public schools. On the one hand, these policies have made it possible for a greater number of young people from low-income classes to pursue higher education, particularly in pedagogy and other teaching degrees. On the other hand, it is a challenge for educational institutions, considering the cultural capital of such students and their families. Many parents have little schooling and many of these students are the first family member to earn a college degree. For these individuals, becoming a teacher means to ascend socially.

162 • LOPES & NACARATO

Finally, it must be pointed out that Brazil does not have public policies for society as a whole; policies are always tied to the interests of whichever political parties are in power. Successful programs are often discarded without further evaluation and others implemented, without consultation of the main players in the educational process, i.e., students and teachers. We are familiar with the discontinuity of public policies, which causes teachers to dismiss them. Many successful innovations and experiences are not even disclosed to teachers and remain restricted to small groups.

Mathematical educators have been presenting proposals for changes in the curricular matrices of teacher education programs. We agree with Moreira (2012), for example, who proposed three questions regarding courses that lead to teaching majors:

1. What kind of mathematics will the teacher teach in elementary school? (know Practice)
2. What mathematics should teachers know to teach that content at school? (design Training). Putting the two questions above into one, we can ask the following question:
3. Is there a way to know mathematics which is specifically appropriate for the professional practice of elementary school teachers? In other words, is there a form of mathematical knowledge which is associated with a professional (educational) view of the mathematics classroom? (pp. 1142–1143)

Regarding teacher education in pedagogy, research has shown that the workload allocated to mathematics for preservice teachers is incompatible with their teaching needs, such as a greater workload of courses focused on specialized knowledge; discussions about teaching and work conditions; and reflective educational practices, enabling the formation of a reflective and critical teacher. It is urgent to review the workload of the disciplines focused on specific content, as well as to reflect on the profile of teacher educators, who must be professionals with mastery of both the epistemological and methodological questions of mathematics teaching.

Concerning more general policies, career plans for teachers must be reviewed in order to make the profession more attractive to young people from different social strata. This requires attention not only to salary issues, but working conditions as well, with fewer students per classroom and better school infrastructure, especially regarding student access to technology.

The education of teachers requires understanding the continuous movement of critical reflection about practice, which, once problematized, with the intention of knowing and understanding its multiple determinations and relations, is also the destination of transformative formative action. Freire points out that: "[...] in the ongoing education of teachers, the fundamental moment is that of critical reflection about practice. By critically thinking about today's or yesterday's practice, can future practice be improved" (1996, p. 39).

Given the expansion of distance education, it is fundamental to analyze the conditions through which such programs are being offered. It is not a question of denying the importance of this mode of delivery in a country the size of Brazil, where, in many regions it is the only possible access to higher education. However, it is not about offering just any program, but programs that are designed to meet the educational demands of children and young people.

We live in uncertain times, due to the disarticulation of many social and educational accomplishments by the new government. The country has already undergone other attempts at the privatization of basic and higher education, at least as far as operations within the school are concerned: control of teacher work by external evaluations, homogenized curricula and market ideology, in a broad and unrestricted adherence to neoliberal principles, similarly to what was denounced by Ravitch (2011) regarding the North American educational system. As Freitas (2014) contends, reformers have accumulated more than 20 years of experience with this neoliberal strand in the United States. Now this experience is being transposed to our country. These are policies of privatization and destabilization of education.

However, even within this uncertain scenario, we have identified successful experiences and resistance movements of mathematics educators who are striving for teacher education focused on social justice, with justice and equality of opportunities and rights for students and preservice teachers. But this is a matter for another text.

REFERENCES

Almeida, N. F. P., Amâncio, M. H., Santos, S. P., & Sales, L. V. (2018). Formação docente e a temática étnico-racial na Revista Brasileira de Educação da Anped (1995–2015) [Teacher education and ethno-racial theme in the Brazilian Journal of Education Anped]. *Revista Brasileira de Educação, 23*, 1–24.

Ball, D. L., Thames, M. H., & Phelps, G. (2008). Content knowledge for teaching: What makes it special? *Journal of Teacher Education, 59*(5), 389–407.

Brazil. (1996). Lei n° 9.394, de 20 de dezembro de 1996. *Estabelece as diretrizes e bases da educação nacional.* [Establishes the guidelines and bases of national education]. Ministério da Educação.

Brazil. (2001). Lei n° 010172, de 9 de Janeiro de 2001. *Aprova o Plano Nacional de Educação e dá outras providências.* [Approves the National Education Plan and determines other measures.]. Ministério da Educação.

Brazil. (2002a). *Resolução CNE/CP 1, de 18 de fevereiro de 2002. Diretrizes Curriculares Nacionais para a Formação de Professores da Educação Básica, em nível superior, curso de licenciatura, de graduação plena* [National Curriculum Guidelines for the Formation of Teachers of the Basic Education, in superior level, degree course]. Ministério da Educação.

Brazil. (2002b). *Decreto n° 4.228*, de 13 de maio de 2002. Institui, no âmbito da Administração Pública Federal, o Programa Nacional de Ações Afirmativas e dá outras providências [It establishes, within the scope of the Federal Public Administration, the National Affirmative Action Program and other measures]. Ministério da Educação.

Brazil. (2004). *Educação inclusiva: A escola* [Inclusive education: The school]. Ministério da Educação, Secretaria de Educação Especial.

Brazil. (2010). Decreto n° 7.352, de 4 de novembro de 2010. *Dispõe sobre a política de educação do campo e o Programa Nacional de Educação na Reforma Agrária—PRONERA*. [Support program for teacher education in rural education.] Ministério da Educação.

Brazil. (2012). Lei n° 12.711, de 29 de agosto de 2012. *Dispõe sobre o ingresso nas universidades federais e nas instituições federais de ensino técnico de nível médio e dá outras providências* [It provides for admission to federal universities and federal high-level technical education institutions and provides other arrangements]. Ministério da Educação.

Brazil. (2015a). Resolução N° 2, de 1° de julho de 2015. *Diretrizes Curriculares Nacionais para a formação inicial em nível superior (cursos de licenciatura, cursos de formação pedagógica para graduados e cursos de segunda licenciatura) e para a formação continuada* [National curriculum guidelines for initial higher education (undergraduate courses, pedagogical training courses for graduates and second degree courses) and for continuing teacher education]. Ministério da Educação.

Brazil. (2015b). *Base Nacional Comum Curricular (BNCC)*. [National Common Curricular Core.] Consulta Pública. MEC/CONSED/UNDIME.

Brazil. (2016a). *Base Nacional Comum Curricular (BNCC). Segunda versão revista.* [National Common Curricular Core, second revision.] MEC/CONSED/UNDIME.

Brazil. (2016b). Portaria CAPES n° 46, de 11 de abril de 2016. *Aprova o Regulamento do Programa Institucional de Bolsa de Iniciação à Docência—Pibid.* [Scholarship Mentorship Initiative Program.] Ministério da Educação (MEC).

Brazil. (2017). Base Nacional Comum Curricular (BNCC). *Educação é a Base.* [National Common Curricular Core.] MEC/CONSED/UNDIME.

Brazil. (2018). Portaria CAPES n° 38, de 28 de fevereiro de 2018. *Institui o Programa de Residência Pedagógica.* [Pedagogical Practicum.] Ministério da Educação (MEC).

Carrillo, J., Climent, N., Contreras, L. C., & Muñoz-Catalán, M. C. (2013). Determining specialized knowledge for mathematics teaching. In B. Ubuz, C. Haser, & M. A. Mariotti (Eds.), *Proceedings of CERME 8* (pp. 2985–2994). METH.

Diniz-Pereira, J. E. (2011). A pesquisa dos educadores como estratégia para construção de modelos críticos de formação docente [Educators' research as a strategy for building critical models of teacher education]. In K. M. Zeichner & J. E. Diniz-Pereira (Org.), *A pesquisa na formação e no trabalho docente* [Research in teacher education and work] (pp. 11–38). Autêntica.

Dourado, L. F., & Oliveira, J. F. (2018). Base Nacional Comum Curricular (BNCC) e os impactos nas políticas de regulação e avaliação da educação superior [Common National Curriculum Base (BNCC) and impacts on higher education regulation and evaluation policies]. In A. da S. Aguiar & L. F. Dourado (Orgs.), *A BNCC na contramão do PNE 2014–2024*: Avaliação e perspectivas [The BNCC against the PNE 2014–2024: Evaluation and perspectives] (pp. 38–43). ANPAE.

Fiorentini, D., Passos, C. L. B., & Lima, R. C. R. (Orgs.) (2016). *Mapeamento da pesquisa acadêmica brasileira sobre o professor que ensina matemática*: Período 2001— 2012 [Survey of Brazilian academic research on the teacher who teaches mathematics: Period 2001—2012]. FE/UNICAMP.

Freire, P. (1996). *Pedagogia da autonomia*: *Saberes necessários à prática educativa* [Pedagogy of autonomy: Knowledge necessary for educational practice]. Paz e Terra.

Freire, P., & Shor, I. (1986). *Medo e ousadia: O cotidiano do professor* [Fearfulness and boldness: The teacher's daily life]. Paz e Terra.

Freitas, L. C. (2014). Os reformadores empresariais da educação e a disputa pelo controle do processo pedagógico na escola [Business reformers of education and the dispute for control of the pedagogical process in the school]. *Educação & Sociedade, 35*(129), 1085–1114.

Garcia, C. M. (1999). *Para uma mudança educativa* [For an educational change]. Porto.

Gatti, B. (2014). A formação inicial de professores para a educação básica: As licenciaturas [Initial teacher education for basic education: Undergraduate degrees]. *Revista USP,* 100, 33–46.

Gatti, B. A., & Barretto, E. S. S. (2009). *Professores do Brasil*: *Impasses e desafios.* [Brazilian teachers: Deadlocks and challenges]. UNESCO.

Gatti, B. A., Barretto, E. S. S., André, M. E. D. A., & Almeida, P. C. A. de A. (2019). *Professores do Brasil: Novos cenários de formação* [Brazilian teachers: New formation scenarios]. UNESCO.

INEP. (Instituto Nacional de Estudos e Pesquisas Educacionais Anísio Teixeira). (2019). *Censo da Educação Superior 2018: notas estatísticas.* [Higher Education Census 2018: statistical notes.]. Ministério da Educação.

Lopes, C. E., & Grando, R. C. (2018). Discussing the mathematics curriculum in Brazil. In D. R. Thompson, M. A. Huntley, & C. Suurtamm (Eds.), *International perspectives on mathematics curriculum* (pp. 189–208). Information Age.

Moreira, P. C. (2012). 3 + 1 e suas (In)Variantes (Reflexões sobre as possibilidades de uma nova estrutura curricular na licenciatura em matemática) [3 + 1 and its invariants (Reflections on the possibilities of a new curricular structure in mathematics degree)]. *BOLEMA*: *Boletim de Educação Matemática (Online)*, *26*, 1137–1150.

Moreira, P. C., & David, M. M. M. S. (2005). O conhecimento matemático do professor: Formação e prática docente na escola básica [Teacher's mathematical knowledge: Teacher education and practice in elementary school]. *Revista Brasileira de Educação, 28*, 50–61.

Moreira, P. C., & Ferreira, A. C. (2013). O lugar da matemática na licenciatura em matemática [The place of mathematics in mathematics degree]. *BOLEMA*: *Boletim de Educação Matemática (Online), 27*, 10–30.

Nóvoa, A. (2007, September). O regresso dos professores [The return of the teachers]. Presented at Conference on Teacher Professional Development for the Quality and Equity of Lifelong Learning, Lisbon, Portugal. http://hdl.handle.net/10451/687

Passos, C. L. B., & Nacarato, A. (2018). Trajetória e perspectivas para o ensino de matemática nos anos iniciais [Trajectory and perspectives for the teaching of mathematics in the early years]. *Estudos Avançados, 94*, 119–135.

Ravitch, D. (2011). *Morte e vida do grande sistema escolar americano* [Death and life of the great American school system]. Ed. Sulina. Brazil.

Santiago, M. E., & Batista Neto, J. (2011). Formação de professores em Paulo Freire: Uma filosofia como jeito de ser-estar e fazer pedagógicos [Teacher education in Paulo Freire: A philosophy as a way of being and teaching]. *Revista e-curriculum*, *7*(3), 1–19.

Shulman, L. S. (1986). Those who understand: knowledge growth in teaching. *Educational Researcher*, *15*(2), 4–14.

Silva, K. A. C. P. da., & Cruz, S. P. (2018). A residência pedagógica na formação de professores: História, hegemonia e resistências. [The pedagogical residence in the formation of teachers: History, hegemony and resistances. Moment: Dialogues in education]. *Momento: diálogos em educação, 27*(2), 227–247.

Universidade Federal do Rio Grande do Sul (UFRGS). (2018). *Projeto Pedagógico de curso: Licenciatura em Pedagogia* [Course Project: Degree in Pedagogy]. Faculdade de Educação. Porto Alegre.

CHAPTER 8

MATHEMATICS TEACHER EDUCATION IN THE UNITED STATES WITH A FOCUS ON INNOVATIONS IN RECRUITMENT AND EQUITABLE INSTRUCTIONAL PRACTICES

Amy Roth McDuffie, Tariq Akmal
Washington State University

Mary Q. Foote
Queens College, City University of New York

In this chapter, we first discuss the context of decentralized teacher education in the U.S. and outline teacher preparation in broad brushstrokes. Next we discuss prominent influences on the vision and focus of Mathematics Teacher Education (MTE) programs, including the complex and multiple professional, governmental, and institutional influences on MTE. This is followed by a more detailed discussion of MTE in national recommendations. We instantiate these ideas by providing two examples of innovative MTE programs in the U.S.—one that addresses the challenge of teacher shortages, and another that prepares teachers to enact equitable practices. We conclude with a discussion of our perspectives on other challenges, innovations, and next steps for MTE in the U.S.

International Perspectives on Mathematics Teacher Education, pages 167–196.
Copyright © 2021 by Information Age Publishing

Describing the structures, goals, visions, designs, and implementations of teacher education programs in the United States (U.S.) is a challenging task, as there is no uniform or common set of policies or legislation that determines how teacher education is provided or how teachers are certified. Thus, perhaps the most appropriate description of mathematics teacher education (MTE) in the U.S. is: "It varies." Although a growing national conversation on teacher preparation is occurring, much of the decision making remains at the state or local level.

THE CONTEXT OF DECENTRALIZED TEACHER EDUCATION

From the earliest days of the U.S., communities desiring a teacher were responsible for developing (or acquiring) their own teachers. As the nation grew, virtually every sizeable population center developed and maintained its own teacher preparation system (Fraser, 2007). This localized, historical control over schools and the preparation of teachers is a theme of education in the U.S. and helps provide a context for the decentralized and disparate influences that buffet teacher preparation. Unlike many nations of the world, individual state legislatures—and not the U.S. government—are the main policy makers and regulators of teacher preparation and education within each state. However, these lines have blurred as the federal government has extended its oversight during the last 30 years, something that has not been well-received by many state education agencies.

For much of the history of the U.S., multiple entry points into the teaching profession resided both within and outside of college or university teacher preparation programs. Approximately two-thirds of certified[1] teachers still enter the field through the approximately 1,400 college or university programs with authorization to prepare teachers. This is described in the next section. However, many other paths to teacher preparation are available that do not include colleges or universities and that contain a wide variety of nonprofit or for-profit entities (National Research Council, 2010). These are discussed later in the chapter.

UNIVERSITY PROGRAMS FOR INTIAL TEACHER PREPARATION: BROAD BRUSHSTROKES

Most U.S. teachers are prepared through a college or university undergraduate baccalaureate program, or through a fifth year Master in Teaching (MiT) or Master of Arts in Teaching (MAT) program. Some states, like California, mainly provide post-baccalaureate teacher preparation programs. Program design varies across colleges and universities. At the elementary level, some colleges and universities offer all content courses for teacher candidates through their education department, and at other places such courses are offered in other departments (e.g., mathematics, English, government, history, music, sciences). Secondary teacher education programs tend to be relatively similar, with candidates typi-

---

[1] In the U.S., some states and professional organizations use other terms for "certificate," including license and endorsement. For consistency, we use the word "certificate" throughout this chapter.

cally following one of two paths; they either earn a degree in a content area (e.g., mathematics, agriculture, biology) and concurrently (or later) complete teacher preparation requirements, or they earn a degree that includes both mathematics content courses and education courses (which often entails specifically prescribed content courses or fewer mathematics courses than required under the other path).

Clinical teaching experiences also vary across programs and states. Clinical placements should be "of sufficient depth, breadth, diversity, coherence, and duration to ensure that candidates demonstrate their developing effectiveness and positive impact on all students' learning and development" (Council for Accreditation of Educator Preparation [CAEP], 2019c, Standard 2.3), but this is often not the case. A required element in each MTE program is supervised student teaching. In some cases, fifth year MiT and MAT programs tend to be free from formal course requirements, and instead include extended clinical experiences. Close relationships between colleges/universities and their local school district partners usually exist, especially with respect to clinical experiences and student teaching.

Near the end of teacher candidates' university teacher education programs, they seek teaching certification through a series of tests. Most states and programs require prospective teachers to pass a basic skills test (or equivalent) and either a performance assessment (edTPA, 2019) or pedagogy test such as the Praxis Principles of Teaching and Learning (PTL; Educational Testing Service [ETS], 2020). In addition, all states require a content knowledge test, using either the Praxis series developed by ETS (2020) or the National Evaluation Series (NES) developed by Pearson (NES, 2020). In a number of states, the NES test is named for the state standards (e.g., Georgia Assessments for the Certification of Educators, Minnesota Teaching Licensure Exam, or Oklahoma Professional Teaching Exam). These exams are discussed in more detail later.

PROMINENT INFLUENCES ON THE VISION AND FOCUS OF MATHEMATICS TEACHER EDUCATION PROGRAMS

Although MTE programs vary across and even within states, the vision and foci for MTE programs is relatively consistent across the U.S. In many ways, this consistency can be attributed to two influences: (1) the release of the Common Core State Standards for Mathematics (CCSSM; National Governors Association Center for Best Practices & Council of Chief State School Officers [NGA & CCSSO], 2010), and (2) policy documents governing MTE and teacher accreditation. All major professional organizations referenced in this chapter and most states have used the CCSSM as a foundational document to inform their standards and recommendations for teacher preparation. Given the importance of the CCSSM in setting a vision and establishing foci for mathematics education and MTE in the U.S., we next provide some history and background on the CCSSM.[2]

---

[2] For more information about CCSSM and grades K–12 education in the U.S., see Remillard and Reinke (2017).

Common Core State Standards for Mathematics

State leaders (e.g., governors and state commissioners of education) led the effort to develop the CCSSM. They sought to establish "consistent, real-world learning goals and launched this effort to ensure all students, regardless of where they live, are graduating high school prepared for college, career, and life" (Common Core State Standards Initiative [CCSSI], n.d. a, para. 1). The CCSSM were informed by existing standards, including *Principles and Standards for School Mathematics* (National Council of Teachers of Mathematics [NCTM], 2000), as well as teachers, mathematicians, university mathematics educators, state leaders, and public feedback. The CCSSM includes two primary sections: *Standards for Mathematical Content* and *Standards for Mathematical Practice*. The *Standards for Mathematical Content* are presented by grade level for kindergarten to Grade 8, and by content area for high school (i.e., number, algebra, functions, modeling, geometry, and statistics and probability). The eight *Standards for Mathematical Practice* apply to all grades, kindergarten through Grade 12, and represent the "processes and proficiencies with longstanding importance in mathematics education" (NGA & CCSSO, 2010, p. 6). As evidence of how the previous NCTM *Standards* (1989, 2000) served as a foundation for the CCSSM, the Standards for Mathematical Practice include the five process standards from NCTM: problem solving, reasoning and proof, communication, representation, and connections. Currently, 41 states have adopted the CCSSM as their curriculum standards, and the CCSSM is highly influential or foundational to the standards in other states (for current information see http://www.corestandards.org/Math).

According to the CCSSI (n.d. b), the CCSSM called for key shifts in practice that have been part of a reform movement in the U.S. over the last three decades (since the *Curriculum and Evaluation Standards* (NCTM, 1989)), with foundational work occurring even earlier. These shifts have influenced recommendations, standards, and programs for teacher preparation in the U.S. Authors of the CCSSM identified shifts in three key areas: focus, coherence, and rigor. These shifts were recommended based on historical U.S. practices of mathematics curriculum and instruction covering too many topics, with too little depth, and with a lack of connections among topics. In addition, U.S. mathematics instruction had tended to focus on procedural learning, with less focus on conceptual understandings and applications of mathematical ideas (*cf.*, Stigler & Hiebert, 1999). Rigor is defined in the CCSSM as follows:

> Rigor refers to deep, authentic command of mathematical concepts, not making math harder or introducing topics at earlier grades. To help students meet the standards, educators will need to pursue, with equal intensity, three aspects of rigor in the major work of each grade: conceptual understanding, procedural skills and fluency, and application. (CCSSI, n.d. b, Shift 3)

TABLE 8.1.   CCSSM Foci by Grand Band

Grade Band	Topic for Deep Focus
Kindergarten to Grade 2	Concepts, skills, and problem solving related to addition and subtraction
Grades 3 to 5	Concepts, skills, and problem solving related to multiplication and division of whole numbers and fractions
Grade 6	Ratios and proportional relationships, and early algebraic expressions and equations
Grade 7	Ratios and proportional relationships, and arithmetic of rational numbers
Grade 8	Linear algebra and linear functions

Building on past standards documents (NCTM, 1989, 2000) and international studies of mathematics curriculum and instruction (*cf.*, Stigler & Hiebert, 1999), recommendations in the CCSSM call for teachers to focus by narrowing and deepening time and effort on mathematics topics. Specifically, teachers at each grade or grade band (through Grade 8) should focus deeply on specific topics, as listed in Table 8.1. These foci are also reflected in standards for teacher preparation (as described in later sections of this chapter).

Accreditation of Teacher Education

As might be expected given the history of state and local control of education in the U.S., states' requirements have differed for the preparation and accreditation of teachers. In 1954, a coalition of educational organizations formed the *National Council for Accreditation of Teacher Education* (NCATE) as a nonprofit, non-governmental body to provide guidance and policy for teacher preparation. Recognized by the U.S. Department of Education as a body to oversee teacher preparation, NCATE developed a set of national standards and entered into agreements with almost all states to adhere to some variation of these standards. Accreditation was not mandatory, however, and only a handful of states *required* NCATE for the review and accreditation of their teacher preparation programs. Predictably, less than half of all educator preparation programs used NCATE. Thus, a set of national standards existed, yet states and universities did not uniformly adopt or follow these standards. In 1997, the *Teacher Education Accreditation Council* (TEAC) was founded with the goal of improving the academic degree programs of professional educators teaching in pre-kindergarten through Grade 12 (CAEP, 2019a). Now there were two competing teacher education accreditation bodies with a wide range of universities and colleges choosing one or the other.

The last 40 years have seen mounting criticism and efforts to disrupt teacher education programs in colleges and universities. Program quality, rigor, diversity, and effectiveness were brought into question (Zeichner, 2018). Alternatives to teacher education programs were developed, and states moved toward allowing other ways of licensing teachers. As states continued to review and approve programs without requiring NCATE or TEAC accreditation, accrediting bodies also felt the pressure of this critique and officials felt that something had to change. Supported by professional organizations, including those in teacher preparation (e.g., American Association of Colleges for Teacher Education, Association of Teacher Educators), state policymakers (e.g., CCSSO, National Association of State Boards of Education), teacher unions, and specialized professional associations (SPAs) such as the Council for Exceptional Children and NCTM, a new vision for merging teacher accreditation took shape. In 2009, the governing boards of NCATE and TEAC approved a joint design team for the *Council for the Accreditation of Education Preparation* (CAEP), a major organization for providing recommendations and accreditation for teacher education programs. In 2013, CAEP became fully operational (CAEP, 2019a).

Like NCATE, CAEP also respected individual state contexts and allowed state officials to regulate teacher education, and hence, provided three options for accreditation: (1) state review only; (2) state and CAEP review with feedback from CAEP; and (3) SPA reviews by professional organizations, such as the NCTM (CAEP, 2019d). In short, state education officials can choose the degree to which their programs align to CAEP Standards. At the time of this writing, all 50 states have agreements with CAEP, but only 16 states have opted for CAEP review with feedback (12 of which require both state and CAEP review), and another 26 use only state-review processes. SPA review is required by 24 states with members in both groups.

CAEP standards for teacher preparation stem from two principles: "solid evidence that the provider's graduates are competent, caring educators;" and "solid evidence that the provider's educator staff have the capacity to create a culture of evidence and use it to maintain and enhance the quality of the professional programs they offer" (2019b, para. 2). CAEP (2019c) uses the following five standards, each with multiple sub-standards: Content and Pedagogical Knowledge; Clinical Partnerships and Practice; Candidate Quality (including state-mandated requirements for testing and minimum grade point averages of 3.0 on a 4.0 scale), Recruitment, and Selectivity; Program Impact; and Provider Quality Assurance and Continuous Improvement.

In 2013, the CCSSO, an organization comprised of the state superintendents in each state, released the Interstate Teacher Assessment and Support Consortium (InTASC) *Model Core Teaching Standards*, which became a required element of the CAEP Standards (2019c). CAEP also required the use of national content standards developed by SPAs (such as NCTM) within all curriculum and assessments. The pedagogical standards of InTASC and the content standards of SPAs were folded into this new accreditation that seemed to reflect all stakeholder requirements.

A DETAILED VIEW OF MATHEMATICS TEACHER PREPARATION IN NATIONAL RECOMMENDATIONS

With this vision and these foci in mind, we now describe U.S. MTE programs in more detail. Given the variability across programs, we discuss standards and recommendations for programs that exist at the national level. It is important to note, however, that these standards are consequential only if state officials and program leaders within states decide to adopt them. Moreover, even if adopted, substantial variability exists in how these standards are enacted in MTE program design and implementation. We first discuss recommendations by professional and accrediting organizations, and then take a detailed look at mathematics teacher preparation by grade band.

Recommendations by Professional and Accrediting Organizations

In Table 8.2 we present a summary of key ideas represented in recommendations and standards for U.S. mathematics teacher preparation programs from major professional and accrediting organizations, including AMTE (2017), Conference Board of the Mathematical Sciences (CBMS, 2010), National Association for the Education of Young Children (NAEYC, 2012), and NCTM and Council for the Accreditation of Educator Preparation (NCTM CAEP, 2012a).[3] The information in these documents is presented differently with varying focus on specific topics; however, across the documents, the ideas and themes are quite consistent and based in research. In generating the table, we drew on AMTE's (2017) *Standards for Preparing Teachers of Mathematics* as the basis for the key components of teacher preparation (shown in the left-hand column in Table 8.2), as it is the most recently published document of those listed above.

A primary recommendation for the preparation of mathematics teachers in the U.S. is as follows:

> Prospective teachers need mathematics courses that develop a solid understanding of the mathematics they will teach. The mathematical knowledge needed by teachers at all levels is substantial yet quite different from that required in other mathematical professions. Prospective teachers need to understand the fundamental principles that underlie school mathematics, so that they can teach it to diverse groups of students as a coherent, reasoned activity and communicate an appreciation of the elegance and power of the subject. Thus, coursework for prospective teachers should examine the mathematics they will teach in depth, from a teacher's perspective. (CBMS, 2010, p. 17)

To meet this goal of preparing teachers to understand deeply the content they will teach, as well as to understand learning and effective pedagogy, teaching certificates in the U.S. are associated with specific grade bands (e.g., early childhood, elementary, middle school, and high school). The range of grades included in specific grade bands for teaching certificates varies by state and by MTE programs. In our descriptions of mathematics teacher preparation requirements, for

---

[3] For mathematics, the CAEP standards are the result of a collaborative effort between CAEP and NCTM (NCTM CAEP, 2012a). NCTM manages the standards and review process for mathematics teacher preparation.

TABLE 8.2. Overview of Recommendations for U.S. Mathematics Teacher Preparation Programs

Areas for Beginning Teachers' Knowledge, Skills, and Dispositions (AMTE, 2017)[a]	Grade Bands			
	Early Childhood Grades PreK–2 (Ages 3–8)	Elementary Grades K–5 (Ages 7–11)	Middle School Grades 6–8 (Ages 12–14)	High School Grades 9–12 (Ages 15–19)
Mathematics Concepts, Practices, and Curriculum	• Counting and cardinality • Operations and algebraic thinking • Number and operations in base 10 • Geometry • Measurement • Data	• Counting and cardinality • Algebraic thinking • Number and operations in base 10 and with fractions and decimals • Geometry • Measurement • Data	• Ratio and proportional relationships • Number systems • Algebraic thinking and functions • Geometry • Measurement • Statistics and probability	• Single and multi-variable calculus • Linear algebra • Statistics and probability • Introduction to proofs • Abstract algebra • Real analysis • Modeling • Graph theory • Number theory • History of mathematics • Analytic geometry • Complex numbers and trigonometry
	Well-prepared beginning teachers of mathematics possess robust knowledge of mathematical and statistical concepts that underlie what they encounter in teaching. They engage in appropriate mathematical and statistical practices and support their students in doing the same. They can read, analyze, and discuss curriculum, assessment, and standards documents as well as students' mathematical productions. (AMTE, 2017 Standard C.1)			
Pedagogical Knowledge and Practices for Teaching Mathematics[b]	Well-prepared beginning teachers of mathematics have foundations of pedagogical knowledge, effective and equitable mathematics teaching practices, and positive and productive dispositions toward teaching mathematics to support students' sense making, understanding, and reasoning. (AMTE, 2017 Standard C.2)			
Students as Learners of Mathematics[b]	Well-prepared beginning teachers of mathematics have foundational understandings of students' mathematical knowledge, skills, and dispositions. They also know how these understandings can contribute to effective teaching and are committed to expanding and deepening their knowledge of students as learners of mathematics. (AMTE, 2017 Standard C.3)			
Social Contexts of Mathematics Teaching and Learning[b]	Well-prepared beginning teachers of mathematics realize that the social, historical, and institutional contexts of mathematics affect teaching and learning and know about and are committed to their critical roles as advocates for each and every student. (AMTE, 2017 Standard C.4)			

[a] The areas listed in this column are from AMTE's (2017) *Standards for Well-Prepared Beginning Teachers of Mathematics.*

[b] For these areas, the knowledge, practices, and dispositions described apply to all grade levels and ages. The descriptions provided represent direct quotations from the AMTE (2017) Standards. Refer to the AMTE Standards for full discussions of each AMTE Standard and corresponding Indicators at https://amte.net/standards.

each grade band we list the core range that is included in most programs. (See Table 8.2, Grade Bands.) It is common for university programs to include overlap among certification ranges for more flexibility in the range of grades that a teacher is certified to teach.

As shown in Table 8.2, mathematics topics often span more than one grade band (e.g., number and operations in base 10) so that the concepts and procedures that are introduced in early grades become more complex in later grades. The mathematics topics listed often correspond to specific course titles (e.g., elementary candidates typically take a course that focuses on number and operations and the base 10 number system); however, the way that university teacher education programs organize courses can vary widely. For example, some universities have a set of courses focused on mathematics content and separate courses focused on pedagogy (often referred to as *methods courses*). Other universities have a series of courses in which content and pedagogy are taught together. At the elementary level, CBMS (2010) recommends that courses blend content and pedagogy as much as possible, but the extent to which universities attend to this recommendation differs. Similarly, high school teacher preparation programs often require separate courses on specific areas of mathematics (e.g., number theory, geometry, mathematical reasoning, and problem solving), as well as a course in the history of mathematics. However, other programs incorporate a focus on the history of mathematics and/or mathematical reasoning and problem solving within other courses. At the middle and high school levels, candidates might take the same courses as those intended for mathematics majors (or a subset of these courses), or they might take specialized versions of these courses that focus on the content needed for future teachers of mathematics, with an emphasis on the coherence and structure of mathematics as it relates to middle or high school course content.

At all grade levels, not only do teachers learn mathematics content, they also learn about the school curriculum, state standards (in most states this is a slightly altered version of CCSSM), and how students learn and develop understandings across grades and within a grade (AMTE, 2017; CBMS, 2010; NCTM CAEP, 2012a). For example, the CCSSM Progressions documents (Institute for Mathematics and Education [IME], n.d.) provide frameworks for a progression of topics within specific mathematics content areas (e.g., Operations and Algebraic Thinking, Fractions, Data) that can support teachers in understanding students' learning by grade. In addition, programs typically offer education courses that focus on topics beyond mathematics (see the portion of Table 8.2 pertaining to Pedagogy, Students, and Social Contexts). These courses have titles such as: *Methods of Teaching Mathematics* (with specific versions for grade bands), *Learning Assessment, Social and Historical Foundations of Education, Human Development, Learning Theory*, and *Diversity in Education*.

In addition to these recommendations for MTE programs, most states require candidates to pass a written exam (or set of exams) in order to receive a teaching certificate. The edTPA exam (2019), which is a portfolio and performance assessment (requiring video of a candidate's teaching), is utilized but not always required for certification in most states in the U.S. Knowledge assessed on this exam is consistent with the information provided in Table 8.2. However, as evidenced in an alignment analysis (NCTM, 2015), the edTPA tends to focus on pedagogy of teaching and learning, rather than mathematics-specific content and pedagogy.

In summary, these documents (i.e., AMTE, 2017; CBMS, 2010; NAEYC, 2012; NCTM CAEP, 2012a) are relatively consistent, yet they each attend to and emphasize different aspects of mathematics teaching and learning. The consistency of recommendations and requirements across professional organizations, accrediting agencies, and assessment systems is due in large part to the fact that all of these teacher preparation documents in the U.S. use the CCSSM as the framework for their recommendations and standards. In addition, these organizations collaborate and build on each other's recommendations and standards to set a common vision and expectations for mathematics teaching and learning in the U.S.

Mathematics Teacher Preparation by Grade Band

One feature of teacher preparation programs that is common across all the grade bands is the importance of including opportunities for candidates to understand research-based learning trajectories and learning progressions (IME, n.d.). Moreover, a positive and supportive learning environment is critical for teaching and learning of all children, regardless of grade or age. A recommendation from AMTE (2017) is that preparation programs should help teacher candidates learn how to create learning environments that focus on exploration, reasoning, and problem solving, and how to draw on children's cultural and linguistic strengths. These positive environments help children develop strong mathematical identities and deep conceptual understanding of mathematics, irrespective of grade level or age. AMTE (2017) recommends that teacher preparation programs support candidates in developing knowledge and practices aimed at cultivating students' mathematics identities and connecting to students' mathematical thinking by building partnerships with families and communities. Teacher candidates also learn how to serve as advocates for students, and how to eliminate institutional barriers to learning.

One example of such a barrier in the U.S. is the common practice of grouping or tracking students by perceived ability. Tracking is a broad term that describes "practices associated with the grouping of students into distinct courses of study" (Domina et al., 2019, p. 293). The practice of tracking communicates "to students the ideas that only some people—particularly white, middle class people—can be good at mathematics" (Boaler, 2011, p. 7). Indeed, based on research, NCTM (2014) calls for eliminating tracking due to the detrimental effects on students, and refers to it as an "obstacle to access and equity" (p. 61). Similar to tracking, ability grouping aims to cluster students by their perceived mathematics ability into homogeneous groups within classrooms (sometimes even at the elementary school level), and is based on the premise that "students have relatively fixed levels of ability and need to be taught accordingly" (Boaler et al., 2000, p. 631). Given that tracking often starts in the middle school grades, we discuss this obstacle further in that section.

We now describe guidelines for mathematics teacher preparation that are specific to the four grade bands outlined in Table 8.2: Early Childhood, Elementary, Middle, and High School. Early Childhood and Childhood (Elementary) Education candidates are prepared to be generalists, teaching all subject areas. For Middle and Adolescent (High School) Education, candidates complete subject-specific programs.

Early Childhood: Pre-Kindergarten–Grade 2, Ages 3–8[4]

In the U.S., educators increasingly are recognizing a need for greater focus on pre-kindergarten, rather than only considering education as starting at kindergarten. This focus is in part due to a range of state policies and practices for preparing early childhood teachers, providing early childhood education programs, and providing systematic support and resources to children and early childhood teachers (CBMS, 2010). Differences in socioeconomic status and associated resources for children are larger in the U.S. than in some other countries (CBMS, 2010). Early childhood learning is critical in developing strong foundations for mathematical thinking and learning in later years (AMTE, 2017).

Despite recommendations at the national level for six to nine semester hours of mathematical content (CBMS, 2010), little guidance is provided about which specific courses should be offered in early childhood teacher preparation programs. In addition to developing deep understandings of early mathematics content, teacher candidates learn about children's development and pedagogical practices for teaching mathematics (AMTE, 2017; NAEYC, 2012). AMTE recommends that preparation programs support teachers in creating learning environments that focus on exploration, reasoning, and problem solving, as well as drawing on children's cultural and linguistic strengths (AMTE, 2017).

In the U.S., many early childhood teachers hold beliefs and understandings that are not supported by current research, including a lack of awareness for what young children can do in mathematics and their potential for mathematics learning, a belief that young children are not ready to learn mathematics, and/or a lack of understanding for how to engage young children in learning mathematics (AMTE, 2017; CBMS, 2010). Teacher preparation programs need to include opportunities for candidates to shift these beliefs and/or develop knowledge and skills to support early learning of mathematics.

---

[4] For some organizations, *early childhood* begins at birth, and for others it begins at age 3. For the purposes of this chapter, we use early childhood to refer to ages 3–8, consistent with AMTE (2017).

Elementary: Kindergarten–Grade 5, Ages 5–11

In the U.S., elementary schools typically include kindergarten to Grade 5, with some schools including Grade 6 and/or pre-kindergarten. Elementary teacher candidates often complete at least 12 semester-hours in courses focused on ideas fundamental to elementary mathematics (as listed in Table 8.2), precursor mathematics for early childhood learners, successor mathematics for middle school students, and pedagogical methods for elementary instruction (CBMS, 2010). Prospective teachers acquire knowledge and skills to support students' learning in upper elementary grades as they build on additive thinking to develop multiplicative thinking, explore fractions and decimals, and operate on rational numbers. Teacher candidates also become skilled at supporting students as they learn about geometric shapes and progress to notice and describe properties of shapes, measurements of shapes, and to develop algebraic thinking (AMTE, 2017). Teacher preparation programs need to provide opportunities for candidates to learn how to support students in developing increasingly sophisticated and abstract mathematical ideas (AMTE, 2017).

Middle School: Grades 6–8, Ages 11–14

A middle grades certificate is offered in 46 states and the District of Columbia; however, many universities offer programs to prepare middle grades mathematics teachers as part of a kindergarten–Grade 8 program or as part of a secondary program (e.g., Grades 6–12) (CBMS, 2010). This grade range tends to vary due to differences in the grades middle schools serve, sometimes even within a given state. For example, a small school district might have one school for kindergarten–Grade 8 and a high school for Grades 9–12. Another district might have elementary schools for kindergarten–Grade 5, middle schools for Grades 6–8, and high schools for grades 9–12. A third structure consists of elementary schools for kindergarten–Grade 6, junior high schools for Grades 7–8, and high schools for Grades 9–12. These three structures are the most common in the U.S., but one can find other structures as well, often based on student enrollments and building capacities within districts (George & Alexander, 2003; Knowles & Brown, 2014). Given this context, teacher education programs tend to have a range of grades such that middle school mathematics teachers are certified to teach in middle school grades that extend beyond the grade range a particular school might serve. Correspondingly, some programs focus only on middle grades, while other programs offer only two ranges: elementary through middle grades (e.g., kindergarten–Grade 8) and secondary grades (e.g., Grades 7–12, Grades 6–12, Grades 5–12). In this section we focus on the areas of preparation aimed specifically at the middle grades (i.e., Grades 6–8).

National guidelines call for prospective middle grades mathematics teachers to complete at least 24 semester-hours of mathematics, including at least 15 semester-hours focused on ideas fundamental to middle school mathematics (CBMS, 2010). AMTE (2017) recommends that prospective teachers learn how to support

adolescent students' ways of thinking and reasoning, as these students often experience cognitive development that exceeds their biological or physical development. Prospective teachers also learn how to implement interdisciplinary tasks
and use instructional strategies that draw on contexts that are culturally relevant,
real-world, and meaningful for adolescent students (AMTE, 2017; CBMS, 2010;
National Middle School Association, 2010). More specifically, teachers are taught
techniques to use students' languages as a resource in learning mathematics as
well as approaches for leveraging students' different cultural experiences and applications of mathematics. This attention to *language and culture as a strength*
provides opportunities for students to learn about and honor their own and other
cultures, as well as build positive mathematics identities.

Prospective middle grades teachers learn strategies to challenge current U.S.
systems and structures that contribute to and perpetuate inequities for early adolescent students in learning mathematics (AMTE, 2017). For example, as mentioned earlier, many schools and districts initiate *tracking* in the middle grades,
and often as early as elementary school (Loveless, 2016). Many educators apply
this practice with good intentions, based on a belief that teachers can better meet
the needs of students of similar ability (Reuman, 1989). However, researchers
have found that tracking creates and perpetuates inequities, especially for African
American, Latinx, and children living in poverty—children in all of these groups
are underrepresented in accelerated tracks (Boaler, 1997, 2008, 2011; Larnell,
2016). Moreover, when teachers and administrators group students by perceived
ability, these groups are often formed based on limited information and a narrow
focus on what counts as ability for mathematics, and thus, strengths and abilities
that contribute to doing mathematics are overlooked and/or undervalued (Boaler,
1997, 2008, 2011).

High School: Grades 9–12, Ages 14–18
In the U.S., high schools typically include Grades 9–12. Similar to the middle
grades, some school districts include a different range of grades in high school or
secondary school (e.g., Grades 10–12, Grades 7–12). We focus here on preparation aimed specifically at the most typical high school range (i.e., Grades 9–12).
As indicated by the topics listed in Table 8.2, CBMS (2010) recommends that prospective high school teachers complete an equivalent of an undergraduate mathematics major that includes at least 9 semester-hours focused on high school mathematics from an advanced viewpoint. Although candidates might take standard
mathematics courses offered by a university, it is recommended that courses either
are adjusted to explicitly connect with high school mathematics or that programs
add other opportunities to gear mathematics content in these courses to the needs
of teacher candidates. The CBMS (2010) and AMTE (2017) documents provide
examples to illustrate how specific topics (e.g., proofs, abstract algebra, modeling,
and differential equations) can be taught in ways that connect key concepts and
processes to high school curriculum and learning. For example, many mathemat-

ics departments offer a course in mathematical modeling (CBMS, 2010). These courses can serve as examples of ways for teacher candidates to incorporate quantitative literacy (e.g., constructing and analyzing statistical models, expressions, equations, or functions) in a real-world context and for a specific purpose. They also provide teacher candidates with opportunities to consider limitations of models and ways to revise models for different purposes (CBMS, 2010). Experiencing these mathematical ideas and processes as learners in college, while considering how the knowledge gained might apply to high school curriculum and learning, is important for teacher candidates' preparation (AMTE, 2017; CBMS, 2010).

By the time students enter high school, they have had substantial opportunities to learn mathematics, both through school and out-of-school experiences. CBMS (2010) recommends that teachers be prepared to support students in building connections between and among mathematics concepts, procedures, strategies, and reasoning processes, as well as to develop productive dispositions toward mathematics. During high school, students will increase their use of tools and technology to learn and do mathematics. Correspondingly, prospective teachers are encouraged to acquire proficiency in using and supporting students to use technology such as spreadsheets, computer algebra systems, dynamic geometry software, statistical analysis, and other tools including physical manipulatives (AMTE, 2017). High school students continue to develop their identities and relationships with mathematics, while considering their future careers and/or education beyond high school. As such, prospective teachers learn skills in supporting students' strengths and confidence to cultivate positive mathematics identities, with particular attention to diverse students' assets and sense-making in mathematics (AMTE, 2017; Celedón-Pattichis et al., 2018). With high school students focusing on next steps beyond high school, prospective teachers are to learn about providing every student with opportunities to think analytically and creatively in the workplace, college, and life beyond school (AMTE, 2017).

TWO EXAMPLES OF INNOVATIVE MATHEMATICS TEACHER EDUCATION PROGRAMS IN THE U.S.

Although it is not possible in this chapter to capture the full range of efforts and innovations to improve teacher preparation programs across the U.S., in this section we discuss two pressing challenges and innovative efforts to address these challenges. The first challenge is recruiting and preparing teachers to address mathematics teacher shortages. The second challenge is preparing teachers to enact equitable and culturally responsive practices.

Addressing Mathematics Teacher Shortages

Recruitment and preparation programs such as the *Teaching Improvements Through Mathematics Education* (TIME) 2000 Program at Queens College, which is part of the City University of New York (CUNY), show promise in addressing mathematics teacher shortages.[5] The TIME 2000 Program is a four-year funded scholarship program in which secondary teacher education candidates are recruited directly from high school. Beginning in 1997, the National Science Foundation (NSF) funded the program for two years. The funding was provided to plan, develop, and implement courses, as well as recruit outstanding mathematics high school students. Students who are recruited are calculus-ready, are recommended by their mathematics teachers, and either have an 85% average or higher in their mathematics courses or have a mathematics SAT score of 600 or above. During the first year, project leaders developed courses and recruited students. In the second year, they offered students full tuition and implemented the courses and seminars. Since the end of NSF funding, other government grants and private foundation funding have been supporting the program (Artzt & Curcio, 2007, 2008; Artzt et al., 2013; Artzt et al., 2012). (See https://www.qc.cuny.edu/Academics/Honors/Time2000 for more information.)

Unlike traditional programs in which teacher candidates often begin their formal teacher preparation coursework in their junior year, TIME 2000 teacher candidates begin their teacher preparation in their first, freshman semester of college. Each year the students form a cohort of approximately 25 students recruited from feeder high schools in the New York Metropolitan area. To recruit students, brochures are sent to schools, announcements are posted on the Internet, current TIME 2000 students return to their high schools to share their experiences, and a special mathematics conference for high school students is hosted by the TIME 2000 Program, with presentations by exemplary mathematics teachers. Although tuition scholarships attract candidates, the reputation of the program—spread by graduates who are in the field teaching—plays a major role in successful recruitment efforts. Students are especially attracted to opportunities the program provides to form close friendships with fellow classmates and with professors who are carefully chosen to teach TIME 2000 courses. Students then enroll in their mathematics and education courses together as a cohort. See Table 8.3 for a list of math courses that students in this program take. This cohort-based scheduling is referred to as *block scheduling*. Although the course titles of the TIME 2000 Program do not always clearly indicate alignment with the content recommended in Table 8.2, the program aligns with AMTE (2017) recommendations. Some of the topics listed in Table 8.2 are included in one or more of the courses listed in Table 8.3 (A. Artzt, personal communication, July 31, 2020).

In addition to coursework, the TIME 2000 program includes field work each semester, either in the form of classroom observations or working directly with students, and culminates in special lesson-study type field work and a student teaching experience. Teacher candidates also meet with each other and the project staff on a monthly basis for seminars, special projects, and events. Topics and

---

[5] Our thanks to Dr. Alice F. Artzt, founder, and Dr. Frances R. Curcio, co-director of the TIME 2000 Program, for their extensive assistance on this section of the chapter.

TABLE 8.3. Mathematics Courses in the TIME 2000 Program

Freshman Year	Sophomore Year	Junior Year	Senior Year
• Differential Calculus • Integral Calculus • Discrete Math	• Calculus II (Infinite Series) • Multivariable Calculus • Probability & Statistics • Methods of Mathematical Statistics • Mathematical Models • Algorithmic Problem Solving (Computer Science)	• Differential Equations • Linear Algebra • Problem Solving • Mathematical Foundations of High School Curriculum • Algebraic Structures	• College Geometry • Foundations of Geometry

events include mathematics and dance, mathematics and art (e.g., Escher art), mathematics and real estate; graduates returning to discuss their lives in the classroom; and undergraduates sharing exemplary mathematics lessons. Candidates participate in individual and small-group conferences with program coordinators for advising and feedback on their performance. In addition, candidates submit monthly journals on their experiences and an annual end-of year e-portfolio in which they document their evolving knowledge, beliefs, and goals with regards to students, mathematics, and teaching. These include supportive artifacts from their course work, seminars, workshops, and conferences that they are required to attend. The program coordinators seek feedback from candidates regarding the program, its strengths and weaknesses, and ways to enhance the program via all of the aforementioned approaches. Candidates also attend at least two mathematics education conferences per year. One of these conferences is hosted by TIME 2000 and takes place at Queens College, and the other is a local mathematics education conference (e.g., Long Island Mathematics Conference).

The TIME 2000 Program reflects a philosophy that prepares teachers to employ approaches that emphasize student-centered instruction, inquiry, and discourse as a means of revealing deep mathematical concepts. For many candidates, these approaches require a shift in traditional beliefs about teaching and learning. By working intensely over an extended period of time (four years) toward program goals, candidates shift their views and practices by:

a. Becoming acclimated to the profession by attending professional conferences, participating in monthly seminars, and experiencing informal mathematics-related trips such as community mathematics walks or visits to New York's National Museum of Mathematics (MoMath);

b. Studying mathematics in a structured and coherent sequence, coordinated with secondary education requirements;

c.   Engaging in field work that incorporates lesson study; and
d.   Participating in innovative, student-centered, problem-based mathematics education courses.

As a testament to the effectiveness of the program, TIME 2000 candidates are specifically sought after and requested to apply for secondary mathematics teaching positions. Furthermore, there is over a 98% passing rate on the required teacher certification examinations, a rate much higher than the statewide average of about 80%. Due to the difficulty level of this program and its stringent requirements, only about 70% of the incoming freshmen who begin the program complete it. However, after graduation, the program boasts an over 95% retention rate during the first five years of teaching compared to 83% nationally. The TIME 2000 program provides a rich example of ways to effectively recruit and prepare teachers to overcome current U.S. teacher shortages.

Preparing Teachers to Enact Equitable and Culturally Responsive Practices

The U.S. has a long history of culturally, linguistically, and socio-economically diverse students in schools. This diversity is coupled with long-standing, institutionalized racism and oppression of non-dominant groups in U.S. educational systems. Although progress has been made in some areas within U.S. educational systems, many educators, students, and families continue to encounter systems of oppression (Frank, 2019; Ladson-Billings, 2005, 2009; Liu & Ball, 2019). The U.S. is not alone in needing to shift to more equitable policies and practices to meet the needs of diverse student populations (*cf.*, de Plevitz, 2007). Although racism can be described from several perspectives, the U.S. Civil Rights Commission described it quite directly as, "Any action or attitude, conscious or unconscious, that subordinated an individual or group based on skin color or race. It can be enacted individually or institutionally" (as cited by Randall, n.d., para. 2). Sensoy and DiAngelo (2017), in describing racism in the U.S., highlighted key aspects relevant for institutions of higher education and teacher preparation programs:

> Racism: White racial and cultural prejudice and discrimination, supported by institutional power and authority, used to the advantage of Whites and the disadvantage of peoples of Color. Racism encompasses economic, political, social, and institutional actions and beliefs that systematize and perpetuate an unequal distribution of privileges, resources, and power between Whites and peoples of Color. (p. 335)

Sleeter (2017) contends that racism represents a way of organizing society and is endemic in U.S. institutions and systems, including education. Within education, she asserts that institutionalized racism influences teacher education in that, "The continued production of teachers, large proportions of whom are not well equipped to teach racially, ethnically, and linguistically diverse students well, is not an aberration. Rather, it is a product of racist systems designed to meet White needs" (p. 157).

Sleeter (2017) further argues that to change these practices and patterns, we need to examine tenets in education, such as claims of neutrality and color-blindness, that can mask White privilege, as well as teacher testing requirements that can perpetuate and reinforce White interests. Daniel Tatum (2017) discusses examples of how color blindness is manifested in U.S. society:

> Color blind racial ideology can be expressed in multiple ways. One is what Ruth Frankenburg calls "color evasion"—as when someone says, "I don't see color; we are all the same," for example. ... Another expression of color blindness is what Frankenburg calls "power evasion," as when someone minimizes the impact of racism, claiming that everyone has the same opportunities to succeed and those who don't have only themselves to blame. (pp. 226–227)

Indeed, at Washington State University and at Queens College, City University of New York, we have found that our teacher candidates often enter our education classes claiming they are "color blind," sincerely believing that this is the best way to approach and support children's learning. In our programs, we work to challenge and shift these views so that teacher candidates understand that a color-blind perspective serves to deny past racism in the U.S., and potentially perpetuates current practices of both explicit and implicit bias and oppression.

In addition, while the U.S. student population is increasingly diverse, the teaching workforce does not reflect the diversity of students they teach: teachers and teacher candidates are predominantly White, female, middle class, and monolingual (Sleeter, 2001, 2017; U.S. Department of Education, 2016). In contrast, current data show that White students no longer comprise a majority in the U.S., and in fact, no single race or ethnic group holds a majority (Hussar & Bailey, 2016;[6] Liu & Ball, 2019).

Teacher candidates are typically underprepared in understanding and meeting diverse students' learning needs (Cruz et al., 2014; Zeichner, 2018), and thus, calls continue for teacher education to better prepare candidates to teach the increasingly diverse student population (Hollins & Guzman, 2005; Nieto, 2000). In mathematics specifically, candidates need support in learning to teach mathematics in ways that support children's understanding to help each child reach their potential (Hiebert et al., 2019; Hiebert et al., 1997). Although preparing teacher candidates to support children's mathematical thinking is critical, in the U.S. effective teaching practices also need to attend to forms of oppression that could serve as barriers to meeting students' needs, such as tracking, holding deficit views toward students and families in non-dominant populations, and school funding policies that instantiate inequities among schools and districts (Celedón-Pattichis et al., 2018). For example, teacher candidates often begin their teacher

---

[6] See Table 7 on p. 47 of the document.

education program with deficit views of children from racially and/or ethnically minoritized groups (Kidd et al., 2008). Deficit perspectives "view students, parents, and communities as lacking in different aspects that enable them to be ready for schooling" (Celedón-Pattichis et al., 2018, p. 375). Many teacher candidates make assumptions about and/or are unaware of others' cultures, and do not understand the role of culture in shaping families' beliefs, values, and child-rearing (Foote et al., 2013; Kidd et al., 2008). Teacher candidates also tend to be unaware of social inequalities associated with race and ethnicity, and this lack of awareness can lead teacher candidates toward deficit perspectives (Kidd et al., 2008).

To meet this challenge, scholars and educators call for a shift from deficit perspectives to asset-based perspectives, and to enact more equitable and culturally responsive practices (Aguirre et al., 2017; Celedón-Pattichis et al., 2018; Rubel, 2017). With an asset-based perspective, we endeavor to connect mathematics to students' diverse cultural, linguistic, and community-based knowledge and experiences in ways that support learning (Gay, 2009; Roth McDuffie, Foote, Bolson et al., 2014; Turner et al., 2012). Specific examples of asset-based approaches in mathematics teacher education courses are described below. National SPAs and other professional organizations (e.g., AMTE, 2017; CBMS, 2010; edTPA, 2019; NCTM, 2014; NCTM CAEP, 2012a) have focused efforts to address these needs and to overcome racist and oppressive policies and practices through calls in current standards, recommendations, and assessments for teaching and teacher preparation. To varying extents, these efforts include a focus on developing more equitable pedagogies to support all students in learning and in developing positive identities towards mathematics. As discussed earlier, this focus is reflected in professional standards and recommendations related to understanding students' thinking and learning with attention to language and culture, developing pedagogical practices that focus on equity, and understanding and incorporating social contexts in teaching and learning.

In our work at Washington State University and Queens College, City University of New York on the *Teachers Empowered to Make Change in Mathematics* project (TEACHMath, a program funded by the National Science Foundation), we developed modules and strategies to prepare prospective teachers to understand and enact equitable practices, recognize resources and assets in all learners, and leverage students' cultural, community, and linguistic backgrounds in teaching and learning mathematics. We refer to students' resources and assets as students' multiple mathematical knowledge bases (MMKB; Turner et al., 2012). For example, in one of the courses that was part of the TEACHMath project, a group of teacher candidates designed a mathematics lesson for a third-grade class with the goal of connecting to students' MMKB (Aguirre et al., 2013). In an effort to learn more about their students' community, the group of teacher candidates visited "Las Socias," a neighborhood food market where Latinx families in the community shopped. The candidates interviewed the market owner about the mathematics she used in her work, and they talked with children and family members about shopping lists, the cost of items, and the money needed each week for groceries. Next, the candidates planned a lesson for which students would calculate the money needed for *Abuela's (Grandma's) shopping list*. This lesson connected to children's and families' experiences shopping for groceries with a reasonable budget in mind. The teacher candidates included items, photos, and prices that reflected items from Las Socias, so the third graders recognized and identified with the context and practices of shopping as part of their own community and experiences (Aguirre et al., 2013). This assignment and other assignments based in the community were designed to re-orient teacher candidates' perspectives to recognize assets in students' communities and the contexts and practices that could be leveraged for teaching mathematics meaningfully (Aguirre et al., 2013; Bartell et al., 2013).

In the TEACHMath project, we also re-designed two commonly used activities in mathematics teacher preparation programs to support teacher candidates in developing asset-based orientations: video case studies and problem-solving interviews. We developed these activities for the elementary mathematics methods courses for teacher candidates, but the project leaders and colleagues have also used them with practicing teachers and in professional development programs (Bartell et al., 2019). For these activities, teacher candidates focus on students' MMKB, on ways to create equitable participation and power structures, and on approaches to learn about and design teaching around students' home and community experiences and cultures. As described by Turner, Foote et al. (2016), for the problem-solving interviews, in addition to posing mathematics problems and asking children about their thinking as they solve the problems, teacher candidates first ask students about their interests and activities, including home and community activities that could include contexts and practices that connect to mathematics, such as sports, flea markets, and doing laundry. Candidates use this information about students' mathematical thinking and knowledge, along with what they know about students' lives and experiences, to develop problems and tasks that can be used to support their students' learning.

As described by Roth McDuffie, Foote, Drake et al. (2014), for the video case studies, teacher candidates view video through four *lenses*. These lenses each have specific prompts concerning teaching, learning, tasks, and power and participation. Although teacher educators often ask candidates to consider teaching, learning, and tasks, each of these lenses includes a prompt to focus candidates' noticing of students' MMKB. For example, the Teaching Lens includes the following prompt:

> What resources and knowledge does the teacher use/draw upon to support students' math understanding (e.g., mathematical, cultural, community, family, linguistic, students' interests, peers)? What does the teacher do to help students make connections within mathematics (e.g., between various mathematical concepts, processes, representations, strategies, etc.)? What does the teacher do to help students connect mathematics with relevant/authentic issues or situations in their lives? What does

the teacher do to encourage students to use their own prior knowledge and/or peers as resources for learning? (p. 137)

In addition to each lens including a prompt that focuses on MMKB, the fourth lens, Power and Participation, focuses teacher candidates' attention on the extent to which all students are encouraged to participate, students hold authority for knowing and learning mathematics, student status and (potentially) differential treatment plays a role in knowing and doing mathematics, student perspectives are valued, and status and power influence learning (Roth McDuffie, Foote, Drake et al., 2014). For additional information about these modules and strategies, see Bartell et al. (2019) and https://TEACHMATH.info.

We conducted case studies of a few of the teacher candidates involved in the TEACHMath elementary mathematics methods courses, following them into student teaching and their first two years of teaching (Turner, Roth McDuffie et al., 2016; Turner et al., 2019). We found that they held asset-based perspectives toward their students and continued to refer to and implement equitable and culturally responsive practices as classroom teachers. Although they would like to go further in implementing these practices, they also reported that they felt limited by pressures and challenges as a novice teacher managing complex work, based on local school and district norms and expectations (e.g., some were in schools with English-only language policies). We expected and often find that early career teachers face multiple challenges as they begin their careers, so finding that the teachers continued to hold asset-based perspectives, along with examples of successfully implementing culturally responsive practices, was encouraging and offers guidance to continuing the practice.

ADDITIONAL CHALLENGES, INNOVATIONS, AND NEXT STEPS

As is evident throughout this chapter, in the U.S. we experience challenges on multiple levels regarding mathematics teacher education. In this section, we focus on four additional challenges that we have identified as particularly pressing: (a) teacher mobility, a lack of coherence in MTE policy and requirements, and reciprocity in teacher education; (b) online and alternative preparation programs; (c) balancing theory and practice in MTE programs; and (d) innovative programs beyond initial teacher preparation to enhance mathematics teacher education.

Teacher Mobility, Lack of Coherence in MTE Policy and Requirements, and Reciprocity in Teacher Education

The U.S. is a mobile society with teachers and families frequently moving among states. Teachers often earn their degrees and/or teaching certificates in one state and then move to another state to work. Moreover, we do not have a nationally centralized policy for teacher education, and we have disparate systems across states and programs. As a result, teachers often earn their teaching certificate, move to another state, and then face new requirements in order to teach. Thus, teachers may need to take additional classes and exams (with fees), or they decide to leave the profession (Dee & Goldhaber, 2017). Moreover, given that certificate grade bands vary by state, teachers could be certified for a grade band in one state (e.g., early childhood or middle grades) and move to another state that does not have that grade band and requires a broader grade range (e.g., elementary or secondary).

In the past, movement from NCATE state to NCATE state was usually straightforward, but not all states were NCATE states. To guide policy on reciprocity, the *National Association of State Directors of Teacher Education and Certification* (NASDTEC) developed the *Interstate Agreement*. This compact (NASDTEC, n.d.) is among U.S. states (and some provinces of Canada) allowing signatories to check credentials of a teacher from another state. The compact also established policies to guide reciprocity of credentials between states without dictating requirements. Although this policy offers solutions among some states, not all states have signed on, so the challenge of mobility still remains for some teachers.

Although it is not likely that the U.S. will convert to a nationalized system for teacher education, organizations that have published recommendations for MTE, such as AMTE (2017) and CBMS (2010), have helped to move MTE toward a common vision, which could support greater coherence. Although no such vision will be perfect, working from a common vision may provide greater opportunities to identify gaps and improve teacher education in the U.S.

Online and Alternative Preparation Programs

Over the last two decades, alternative routes to certification have emerged that offer more flexible approaches to earning teacher certification. For example, a rapidly increasing number of university-based online teacher preparation programs are now available nationally for those who cannot access traditional teacher preparation (e.g., due to schedules or distance). Regionally accredited online programs meet the same requirements as other preparation programs in each state and some programs also opt for national accreditation. Notable online providers for initial certification include Arizona State University, Grand Canyon University, Walden University, and Western Governors University (a consortium of 19 states). Teacher candidates take courses in online programs that purport to meet all state requirements for certification. Clinical placements may be the candidate's responsibility, which may be a challenge for the candidate to negotiate, particularly in larger districts with centralized placement systems; also, candidates may need to figure out who a good mentor teacher may be (Zeichner, 2018). In addition, field supervisors traditionally are advocates for the teacher candidate as evaluators. Having a placement without a supervisor or a district-based supervisor may limit the level of support a candidate receives.

According to Zeichner (2018), amid recent teacher shortages, some state education officials have incentivized or required faculty at colleges and universities to create alternative routes to teacher certification to serve students who are unable to access traditional programs. These routes vary greatly across states and allow for accelerated pathways and credit for prior learning experiences. These programs are often competency, rather than credit-based, approaches. Such programs often vary their delivery mechanisms by including online courses and/or offering courses during evening hours, on weekends, or during summer months. Unfortunately, programs independent of colleges and universities have also appeared and are funded by venture capitalists or corporations run by people who seek profitability. Examples include unaccredited charter schools that offer teacher certification, some of which are now applying for accreditation. Many of these alternative certification route programs boast short time frames, early entry to classrooms (candidates are teachers of record having had very little coursework), are entirely taught by non-teacher education scholars, and often serve students who come to them with a philosophy that matches that of the program. In haste to open as many doors as possible, many state education officials are approving such programs that offer experiences running counter to what is required by state accreditation agencies and are not grounded in research-based practices. Alternate route and non-university programs can differ substantially from university-based programs and from each other, and they are often criticized for lacking quality and rigor, as well as program officials being questioned about their intentions.

Balancing Theory and Practice in MTE Programs

Although we have described sets of standards and requirements, along with visions for teacher preparation programs, these descriptions do not tend to capture underlying conceptual framings for meaningful learning in mathematics teacher education. Indeed, scholars and teacher educators contend that a balance between current research and theory, and practice-based knowledge and skills, is needed to effectively prepare teachers (*cf.*, Putnam & Borko, 2000). In the U.S., MTE programs, partnering schools, and mentor teachers are faced with the challenge of providing an appropriate balance of theory and practice, and determining which experiences are most effective in supporting teacher candidates' learning and development, as well as practicing teachers' on-going professional development. Many educators and researchers are working toward developing courses and programs that achieve this balance.

For example, the Cognitively Guided Instruction Project (CGI; Carpenter et al., 2015) was a U.S.-based NSF-funded research project that began in the 1990s and continues to impact teacher education today. CGI researchers studied how elementary students' thinking and learning related to the basic mathematics operations (addition, subtraction, multiplication, and division), and created a framework of problem types and strategies children use to solve these types of problems. CGI researchers then created a professional development program designed to inform elementary teachers about children's mathematical thinking and problem solving. They examined student achievement in classes taught by CGI-trained teachers, as well as how teachers changed their instruction after participating in the CGI professional development (*cf.*, Carpenter et al., 1989; Fennema et al., 1996). CGI research is typically part of elementary mathematics methods courses and is also featured in elementary mathematics textbooks (*cf.*, Van de Walle et al., 2016). Moreover, mathematics teacher educators often create classroom-based assignments in which candidates interview and/or observe students solving CGI problem types so that candidates develop research-based understandings about children's mathematical thinking (*cf.*, Bartell et al., 2019). The CGI project illustrates an effective way that theory has been used to inform practice, teacher education, and professional development.

Innovative Programs Beyond Initial Teacher Preparation to Enhance Mathematics Teacher Education

Although this chapter focuses on initial teacher preparation, in recent years, programs have been designed to support practicing teachers in developing their expertise and becoming accomplished teachers. In this section, three such programs are described.

Many states offer a certification for elementary mathematics specialists (EMS), which is an advanced certification for experienced teachers. Although it is not an official certification in all U.S. states, almost all states call on experienced expert teachers to support mathematics teaching, especially in the elementary grades (AMTE, 2013). Currently, 19 states offer some form of elementary mathematics specialist certification, and an additional nine states are in the process of adding this certification (Elementary Mathematics Specialists & Teacher Leaders Project, 2019). In the U.S., most elementary teachers are generalists, responsible for teaching all subject areas. Given the breadth of knowledge needed, the EMS position and certification are intended to support elementary teachers in building and enacting deeper understandings of mathematics teaching and learning (NCTM, 2010). AMTE originally developed *Standards for Elementary Specialists* in 2010, and then updated these standards in 2013, based on the CCSSM. Similar to other standards and recommendations, these standards were developed based on and with input from other major professional organizations, including CBMS (2010); later CAEP adopted these standards for accrediting programs that offer an EMS certificate (NCTM CAEP, 2012b). Although the role of EMS can vary widely in different schools, districts, and states, and these specialists perform tasks that range from providing classroom-based support to state-level leadership, at its root, EMS support effective mathematics instruction. EMS may work with students in classrooms, coach teachers in their classrooms, and/or provide professional development with a focus on school-, district-, or state-wide improvement (AMTE, 2013). Research on schools and districts with professionals serving in EMS roles has shown positive effects on teachers' development and students' learning (*c.f.*, Campbell et al., 2013; Campbell et al., 2014; Kessel, 2009).

A second form of advancement for experienced teachers is National Board Certification (NBC). Teachers must have completed three years of classroom teaching prior to engaging in the NBC process. NBC is available in 25 certificate areas, including *Early Adolescent Mathematics* (ages 11 to 15) and *Adolescent and Young Adult Mathematics* (ages 14 to 18+) (National Board for Professional Teaching Standards, n.d.). The National Board program requires candidates seeking NBC to pass a computer-based written assessment and a portfolio assessment (including video of teacher candidates' teaching) that demonstrates they meet all National Board Standards. The National Board Standards are consistent with recommendations from other professional organizations described in this chapter, with the additional focus on professional leadership.

A third program for experienced teachers, which shows great promise, is the New York State Master Teacher Program. This program is a professional network of teachers comprised of over 800 public school teachers throughout New York State and focuses on developing teaching in STEM[7] fields and computer science. Professional learning for these teachers is administered through nine State University of New York (SUNY) campuses. Teachers engage in work in the areas of (a) knowledge of STEM content, (b) knowledge of pedagogy, and (c) knowledge of students and their families and communities. Candidates for the program must be certified grades K–12 public school teachers with at least four years of experience (SUNY, 2020).

CONCLUSION

We face many challenges in the U.S. regarding mathematics teacher education, such as tremendous variation in teacher preparation programs within and across states. At the same time, we also see great promise. Organizations such as AMTE and NCTM are leading initiatives to critically examine and improve MTE programs and mathematics teaching. In addition, the NSF has made substantial investments in MTE programs and professional development, as well as research on the teaching and learning of mathematics. Through efforts and investments such as these, we hope that we can take on these challenges and improve mathematics teaching and children's learning in the future.

REFERENCES

Aguirre, J., Herbel-Eisenmann, B., Celedón-Pattichis, S., Civil, M., Wilkerson, T., Stephan, M., & Clements, D. (2017). Equity within educational research as a political act. Moving from choice to intentional collective professional responsibility. *Journal for Research in Mathematics Education, 48*(2), 124–147.

Aguirre, J., Turner, E., Bartell, T. G., Kalinec-Craig, C., Foote, M. Q., Roth McDuffie, A., & Drake, C. (2013). Making connections in practice: How prospective elementary teachers connect to children's mathematical thinking and community funds of knowledge in mathematics instruction. *Journal of Teacher Education, 64*(2), 178–192. doi:10.1177/0022487112466900

---

[7] STEM: science, technology, engineering, and mathematics

Artzt, A. F., & Curcio, F. R. (2007). TIME 2000: A mathematics teaching program. *Mathematics Teacher, 100*, 542–543.

Artzt, A. F., & Curcio, F. R. (2008). Recruiting and retaining secondary mathematics teachers: Lessons learned from an innovative, four-year, undergraduate program. *Journal of Mathematics Teacher Education, 11*(3), 243–251.

Artzt, A. F., Curcio, F. R., & Sultan, A. (2013, April). Queens College: A program for math teachers requires a complex formula. *Phi Delta Kappan, 94*(7), 23.

Artzt, A. F., Sultan, A., Curcio, F. R., & Gurl, T. (2012). A capstone mathematics course for prospective secondary mathematics teachers. *Journal of Mathematics Teacher Education, 15*, 251–262.

Association of Mathematics Teacher Educators (AMTE). (2013). *Standards for elementary mathematics specialists: A reference for teacher credentialing and degree programs.* https://amte.net/sites/all/themes/amte/resources/EMS_Standards_AMTE2013.pdf

Association of Mathematics Teacher Educators. (2017). *Standards for preparing teachers of mathematics.* https://amte.net/standards

Bartell, T. G., Drake, C., Roth McDuffie, A., Aguirre, J. M., Turner, E. E., & Foote, M. Q. (2019). *Transforming MTE: An equity-based approach.* Springer.

Bartell, T. G., Foote, M. Q., Drake, C., Roth McDuffie, A., Turner, E. E., & Aguirre, J. M. (2013). Developing teachers of Black children: (Re)orienting thinking in an elementary mathematics methods course. In J. Leonard & D. B. Martin (Eds.), *The brilliance of Black children in mathematics: Beyond the numbers and toward a new discourse* (pp. 343–367). Information Age Publishing.

Boaler, J. (1997). When even the winners are losers: Evaluating the experiences of top set students. *Journal of Curriculum Studies, 29*(2), 165–182.

Boaler, J. (2008). *What's math got to do with it?: How parents and teachers can help children learn to love their least favorite subject.* Penguin.

Boaler, J. (2011). Changing students' lives through the de-tracking of urban mathematics classrooms. *Journal of Urban Mathematics Education, 4*(1), 7–14.

Boaler, J., Wiliam, D., & Brown, M. (2000). Students' experiences of ability grouping-disaffection, polarisation and the construction of failure. *British Educational Research Journal, 26*(5), 631–648. https://doi.org/10.1080/713651583

Campbell, P. F., Ellington, A. J., Haver, W. E., & Inge, V. L. (2013). *Handbook for elementary mathematics specialists.* National Council of Teachers of Mathematics.

Campbell, P. F., Nishio, M., Smith, T. M., Clark, L. M., Conant, D. L., Rust, A. H., DePiper, J. N., Frank, T. J., Griffin, M. J., & Choi, Y. (2014). The relationship between teachers' mathematical content and pedagogical knowledge, teachers' perceptions, and student achievement. *Journal for Research in Mathematics Education, 45*, 419–459.

Carpenter, T., Fennema, E., Franke, M., Levi, L., & Empson, S. (2015). *Children's mathematics: Cognitively guided instruction* (2nd Ed.). Heinemann.

Carpenter, T., Fennema, E., Peterson, P., Chiang, C., & Loef, M. (1989). Using knowledge of children's mathematics thinking in classroom teaching: An experimental study. *American Educational Research Journal, 26*(4), 499–531.

Celedón-Pattichis, S., Peters, S., Borden, L., Males, J., Pape, S., Chapman, O., Clements, D., & Leonard, J. (2018). Asset-based approaches to equitable mathematics education research and practice. *Journal for Research in Mathematics Education, 49*, 373–389.

Common Core State Standards Initiative. (n.d.a). *Development process*. http://www.cores-tandards.org/about-the-standards/development-process/

Common Core State Standards Initiative. (n.d.b). *Key shifts in mathematics*. http://www.corestandards.org/other-resources/key-shifts-in-mathematics/

Conference Board of the Mathematical Sciences (CBMS). (2010). *The mathematical education of teachers II*. American Mathematical Society.

Council for Accreditation of Education Preparation (CAEP). (2019a). *History of CAEP*. http://caepnet.org/about/history

Council for Accreditation of Education Preparation. (2019b). *Introduction to the standards*. http://caepnet.org/standards/introduction

Council for Accreditation of Education Preparation. (2019c). *2013 CAEP standards*. http://caepnet.org/~/media/Files/caep/standards/caep-standards-one-pager-0219.pdf?la=en

Council for Accreditation of Education Preparation (2019d). *State partnership agreements*. http://caepnet.org/working-together/state-partners

Cruz, B. C., Ellerbrock, C. R., Vasquez, A., & Howes, E. V. (2014). *Talking diversity with teachers and teacher educators: Exercises and critical conversations across the curriculum*. Teachers College Press.

Daniel Tatum, B. (2017). *Why are all the black kids sitting together in the cafeteria? And other conversations about race*. Basic Books.

Dee, T., & Goldhaber, D. (2017). *Understanding and addressing teacher shortages in the United States: The Hamilton Project*. Brookings. https://www.brookings.edu/re-search/understanding-and-addressing-teacher-shortages-in-the-united-states/

de Plevitz, L. (2007). Systemic racism: The hidden barrier to success for Indigenous school students. *Australian Journal of Education, 51*(1), 54–71.

Domina, T., McEachin, A., Hanselman, P., Agarwal, P., Hwang, N., & Lewis, R. W. (2019). Beyond tracking and detracking: The dimensions of organizational differentiation in schools. *Sociology of Education, 92*(3), 293–322.

edTPA. (2019). *edTPA*. Pearson. http://www.edtpa.com/

Educational Testing Service (ETS). (2020). *ETS Praxis*. https://www.ets.org/praxis/

Elementary Mathematics Specialists & Teacher Leaders Project. (2019). *Elementary mathematics specialists and teacher leaders project*. http://www.mathspecialists.org/

Fennema, E., Carpenter, T., Franke, M., Levi, L., Jacobs, V., & Empson, S. (1996). Learning to use children's thinking in mathematics instruction: A longitudinal study. *Journal for Research in Mathematics Education, 27*(4), 403–434.

Foote, M. Q., Roth McDuffie, A., Turner, E. E., Aguirre, J. M., Bartell, T. G., & Drake, C. (2013). Orientations of prospective teachers towards students' families and communities. *Teaching and Teacher Education, 35*, 126–136.

Frank, T. J. (2019). Using critical race theory to unpack the Black mathematics teacher pipeline. In J. Davis & C. Jett (Eds.), *Critical race theory in mathematics education* (pp. 98–122). Routledge.

Fraser, J. (2007). *Preparing America's teachers: A history*. Teacher's College Press.

Gay, G. (2009). Preparing culturally responsive mathematics teachers. In B. Greer, S. Muk-hopadhyay, A. B. Powell, & S. Nelson-Barber (Eds.), *Culturally responsive mathematics education* (pp. 189–205). Routledge.

George, P. S., & Alexander, W. M. (2003). *The exemplary middle school* (3rd Ed.). Wadsworth Publishing.

Hiebert, J., Berk, D., Miller, E., Gallivan, H., & Meikle, E. (2019). Relationships between opportunity to learn mathematics in teacher preparation and graduates' knowledge for teaching mathematics. *Journal for Research in Mathematics Education, 50*(1), 23–50.

Hiebert, J., Carpenter, T. P., Fennema, E., Fuson, K. C., Wearne, D., Murray, H., & Human, P. (1997). *Making sense: Teaching and learning mathematics with understanding.* Heinemann.

Hollins, E., & Guzman, M. T. (2005). Research on preparing teachers for diverse populations. In M. Cochran-Smith & K. Zeichner (Eds.), *Studying teacher education: The report of the AERA panel on research and teacher education* (pp. 477–548). Lawrence Erlbaum Associates.

Hussar, W. J., & Bailey, T. M. (2016). *Projections of educational statistics to 2023* (NCES 2015-073). U.S. Department of Education, National Center for Educational Statistics. U.S. Government Printing Office.

Institute for Mathematics and Education. (n.d.). *Progressions documents for the Common Core math standards.* http://math.arizona.edu/~ime/progressions/

Kessel, C. (Ed.). (2009). *Teaching mathematics: Research, ideas, projects, evaluation.* http://www.msri.org/calendar/attachments/workshops/430/TTM_EdSeries3MSRI.pdf

Kidd, J. K., Sanchéz, S. Y., & Thorp, E. K. (2008). Defining moments: Developing culturally responsive dispositions and teaching practices in early childhood preservice teachers. *Teaching and Teacher Education, 24*, 316–329.

Knowles, T., & Brown, D. F. (2014). *What every middle school teacher should know* (3rd Ed.). Heinemann.

Ladson-Billings, G. (2005). Is the team all right? Diversity and teacher education. *Journal of Teacher Education, 56*(3), 229–234.

Ladson-Billings, G. (2009). Race still matters: Critical race theory in education. In M. S. Apple, W. Au, & L. A. Gandin (Eds.), *The Routledge international handbook of critical education* (pp. 110–122). Routledge.

Larnell, G. V. (2016). More than just skill: Examining mathematics identities, racialized narratives, and remediation among black undergraduates. *Journal for Research in Mathematics Education, 47*(3), 233–269.

Liu, K., & Ball, A. (2019). Critical reflection and generativity: Toward a framework of transformative education for diverse learners. In T. Pigott, A. M. Ryan, & C. Tocci (Eds.), *Review of research in education: Changing teaching practice in P–20 educational settings* (Vol. 43, pp. 68–105). American Educational Research Association and Sage.

Loveless, T. (2016). *How well are American students learning? Part II: Tracking and advanced placement. 3*(5), 16–25. http://www.brookings.edu/~/media/Research/Files/Reports/2016/03/brown-center-report/Brown-Center-Report-2016.pdf?la=en

National Association for the Education of Young Children (NAEYC). (2012). *2010 NAEYC standards for initial & advanced early childhood professional preparation programs.* http://www.naeyc.org/caep/files/caep/NAEYC%20Initial%20and%20Advanced%20Standards%2010_2012.pdf

National Association of State Directors of Teacher Education and Certification (NASDTEC). (n.d.). *Interstate agreement.* https://www.nasdtec.net/page/Interstate

National Board for Professional Teaching Standards. (n.d.). *National Board Certification overview*. https://www.nbpts.org/national-board-certification/overview/

National Council of Teachers of Mathematics. (1989). *Curriculum and evaluation standards for school mathematics*. Author.

National Council of Teachers of Mathematics. (2000). *Principles and standards for school mathematics*. Author.

National Council of Teachers of Mathematics. (2010). *The role of elementary mathematics specialists in teaching and learning mathematics*. https://www.nctm.org/Standards-and-Positions/Position-Statements/The-Role-of-Elementary-Mathematics-Specialists-in-the-Teaching-and-Learning-of-Mathematics/

National Council of Teachers of Mathematics. (2014). *Principles to actions: Ensuring mathematical success for all*. Author.

National Council of Teachers of Mathematics. (2015). *Alignment of NCTM CAEP Standards (2012) for Secondary to edTPA rubrics*. https://www.nctm.org/Standards-and-Positions/CAEP-Standards/

National Council of Teachers of Mathematics Council for the Accreditation of Educator Preparation. (2012a). *CAEP Standards for mathematics teacher preparation*. https://www.nctm.org/Standards-and-Positions/CAEP-Standards/

National Council of Teachers of Mathematics Council for the Accreditation of Educator Preparation. (2012b). *NCTM CAEP Standard for elementary school specialists*. https://www.nctm.org/Standards-and-Positions/CAEP-Standards/

National Evaluation Series (NES). (2020). NES. Pearson Education. https://www.nestest.com/

National Governors Association Center for Best Practices & Council of Chief State School Officers. (2010). *Common core state standards for mathematics*. Author. http://www.corestandards.org/Math

National Middle School Association. (2010). *This we believe: Keys to educating young adolescents* (4th Ed.). National Middle School Association (NMSA, now Association for Middle Level Education).

National Research Council. (2010). *Preparing teachers: Building evidence for sound policy*. The National Academies Press. https://doi.org/10.17226/12882

Nieto, S. (2000). Placing equity front and center: Some thoughts on transforming teacher education for a new century. *Journal of Teacher Education, 51*(3), 180–187.

Putnam, R., & Borko, H. (2000). What do new views of knowledge and thinking have to say about research on teacher learning? *Educational Researcher, 29*(1), 4–15.

Randall, V. (n.d.). *Defining race, racism, and racial discrimination*. https://racism.org/articles/race/what-is-race/276-race0801

Remillard, J., & Reinke, L. (2017). Mathematics curriculum in the United States: New challenges and opportunities. In D. R. Thompson, M. A. Huntley, & C. Suurtamm (Eds.), *International perspectives on mathematics curriculum* (pp. 131–162). Information Age Publishing.

Reuman, D. A. (1989). How social comparison mediates the relation between ability-grouping practices and students' achievement expectancies in mathematics. *Journal of Educational Psychology, 81*(2), 178–189. https://doi.org/10.1037/0022-0663.81.2.178

Roth McDuffie, A., Foote, M. Q., Bolson, C., Turner, E. E., Aguirre, J. M., Bartell, T. G., Drake, C., & Land, T. (2014). Using video analysis to support prospective K–8 teachers' noticing of students' multiple mathematical knowledge bases. *Journal of Mathematics Teacher Education, 17*, 245–270. doi:10.1007/s10857-013-9257-0

Roth McDuffie, A., Foote, M. Q., Drake, C., Turner, E. E., Aguirre, J. M., & Bartell, T. G. (2014). Mathematics teacher educators' use of video analysis to support prospective K–8 teachers' noticing. *Mathematics Teacher Educator, 2*(2), 108–140.

Rubel, L. (2017). Equity-directed instructional practices: Beyond the dominant perspective. *Journal of Urban Mathematics Education, 10*(2), 66–105.

Sensoy, Ö. J., & DiAngelo, R. J. (2017). *Is everyone really equal?: An introduction to key concepts in social justice education.* Teachers College Press.

Sleeter, C. E. (2001). Preparing teachers for culturally diverse schools: Research and the overwhelming presence of whiteness. *Journal of Teacher Education, 52*(2), 94–106.

Sleeter, C. E. (2017). Critical race theory and the Whiteness of teacher education. *Urban Education, 53*(2), 155–169.

State University of New York. (2020). *About the master teacher program.* https://www.suny.edu/masterteacher/about/

Stigler, J., & Hiebert, J. (1999). *The teaching gap: Best ideas from the world's teachers for improving education in the classroom.* Simon and Schuster.

Turner, E., Drake, C., Roth McDuffie, A., Aguirre, J., Bartell, T., & Foote, M. (2012). Promoting equity in mathematics teacher preparation: A framework for advancing teacher learning of children's multiple mathematics knowledge bases. *Journal of Mathematics Teacher Education, 15*(1), 67–82. doi:10.1007/s01857-011-9196-6

Turner, E., Roth McDuffie, A., Sugimoto, A., Aguirre, J., Bartell, T. G., Drake, C., Foote, M., Stoehr, K., & Witters, A. (2019). A study of early career teachers' practices related to language and diversity during mathematics instruction. *Mathematical Thinking and Learning, 21*(1), 1–26. doi:10.1080/10986065.2019.1564967

Turner, E., Roth McDuffie, A., Sugimoto, A., Stoehr, K., Witters, A., Aguirre, J., Bartell, T. G., Drake, C., & Foote, M. (2016). Early career elementary mathematics teachers' noticing related to language and language learners. In M. Wood, E. Turner, & M. Civil (Eds.), *Proceedings of the 38th Annual Meeting of the North American Chapter for the Psychology of Mathematics Education* (pp. 347–354). University of Arizona.

Turner, E. E., Foote, M. Q., Stoehr, K. J., Roth McDuffie, A., Aguirre, J. M., Bartell, T. G., & Drake, C. (2016). Learning to leverage children's multiple mathematical knowledge bases in instruction. *Journal of Urban Mathematics Education, 9*(1), 48–78.

U.S. Department of Education. (2016). *The state of racial diversity in the educator workforce.* http://www2.ed.gov/rschstat/eval/highered/racial-diversity/state-racial-diversityworkforce.pdf

Van de Walle, J., Karp, K., & Bay-Williams, J. (2016). *Elementary and middle school mathematics: Teaching developmentally* (9th Ed.). Longman.

Zeichner, K. M. (2018). *The struggle for the soul of teacher education.* Routledge.

CHAPTER 9

MATHEMATICS TEACHER EDUCATION IN CANADA

Emerging Possibilities in Engaging with Indigenous Perspectives

Florence Glanfield
University of Alberta

Christopher Charles
University of Manitoba

This chapter outlines the ways in which the engagement of Indigenous[1] perspectives in mathematics teacher education have unfolded in Canada over the last 20 years. A number of different policies and reports have dramatically changed the Canadian landscape of mathematics teacher education so that teacher education programs are now engaging with Indigenous perspectives. The chapter starts by providing the context of K–grade 12 education and initial teacher education and certification in Canada, followed by a description of the policies and reports that have shaped the current landscape. We then look at one example, as a case, of the ways in which a secondary mathematics teacher education program in Canada is beginning to take up the challenge of engaging with Indigenous perspectives to illuminate the complexities of implementing these perspectives in a mathematics teacher education program.

---

[1] Indigenous, or Aboriginal, peoples of Canada is a collective name for the original peoples and their descendants. According to the Canadian constitution (1982), there are three recognized groups of Aboriginal peoples in Canada: First Nations, Inuit, and Métis (Government of Canada, 2020a).

International Perspectives on Mathematics Teacher Education, pages 197–213.

EDUCATION IN CANADA

The British North America Act, a law passed by the British parliament in 1867 to form Canada, outlined responsibilities for education. Section 93 of the Act states, "In and for each Province the Legislature may exclusively make Laws in relation to Education…" (Government of Canada, 2020b, para. 1). The current context is such that each of the 10 provinces and 3 territories has legislation related to K–12 education, the teaching profession, and post-secondary education. In each of the 13 provinces and territories, there is a Ministry of Education that is responsible for K–12 education. In most provinces there is also a Ministry of Post-Secondary Education. Elementary and secondary public education is provided free to all Canadian children and youth[2] (Council of Ministers of Education, Canada, 2019a).

Standards for initial and ongoing teacher certification in Canada fall under Section 93 of the British North America Act. Hence, each province and territory has its own legislation related to teacher certification. Universities, where initial teacher education is conducted, generally fall under the province's Ministry of Post-Secondary Education; when there is no Ministry of Post-Secondary Education, then teacher education is included in the Ministry of Education. There are 64 university faculties,[3] colleges, schools, and departments in Canada that offer initial teacher education leading to initial teacher certification. Graduates of initial teacher education programs are granted initial teacher certification in their respective province upon successful completion of their university teacher education programs.[4] Every initial teacher education program is required to meet the standards for initial teacher certification outlined by their respective provincial/territorial legislation. Every teacher in Canada who is certified by a province/territory has at least 4 years of undergraduate education and must demonstrate fluency in at least one of the two official languages—English or French.

Although there are 13 different educational jurisdictions in Canada, there are two national organizations that discuss teacher certification and teacher education. One organization is the Council of Ministers of Education, Canada (CMEC). CMEC is an intergovernmental body founded in 1967 by ministers of education to serve as "a forum to discuss policy issues; a mechanism through which to undertake activities, projects, and initiatives in areas of mutual interest; an instrument to represent the education interests of the provinces and territories internationally" (CMEC, 2019a, para. 1).

A second organization is the Association of Canadian Deans of Education (ACDE), a group affiliated with the Canadian Society for the Study of Education (CSSE) and comprising the Deans, Directors, and Chairs of the 64 initial teacher

---

[2] See Simmt (2018) for more information about K–12 mathematics curriculum in Canada.

[3] In Canadian universities, the word "faculty" often refers to an organizational unit, similar to a department. In this chapter, "faculty" often refers to a Faculty of Education.

[4] If a teacher has certification in one province or territory, then the teacher is eligible for certification in other provinces or territories in Canada (Council of Ministers of Education, Canada [CMEC], 2020).

An Effective Initial Teacher Education Program

- Demonstrates the transformative power of learning for individuals and communities.
- Envisions the teacher as a professional who observes, discerns, critiques, assesses, and acts accordingly.
- Encourages teachers to assume a social and political leadership role.
- Cultivates a sense of the teacher as responsive and responsible to learners, schools, colleagues, and communities.
- Involves partnerships between the university and schools, interweaving theory, research, and practice and providing opportunities for teacher candidates to collaborate with teachers to develop effective teaching practices.
- Promotes diversity, inclusion, understanding, acceptance, and social responsibility in continuing dialogue with local, national, and global communities.
- Engages teachers with the politics of identity and difference and prepares them to develop and enact inclusive curricula and pedagogies.
- Supports a research disposition and climate that recognizes a range of knowledge and perspectives.
- Ensures that beginning teachers understand the development of children and youth (intellectual, physical, emotional, social, creative, spiritual, moral) and the nature of learning.
- Ensures that beginning teachers have sound knowledge of subject matter, literacies, ways of knowing, and pedagogical expertise.
- Provides opportunities for candidates to investigate their practices.
- Supports thoughtful, considered, and deliberate innovation to improve and strengthen the preparation of educators.

FIGURE 9.1. Principles for Initial Teacher Education Programs in Canada (Association of Canadian Deans of Education, 2016, p. 4)

education programs across Canada. ACDE is "committed to pan-Canadian leadership in professional and teacher education, educational research, and policy in universities and university-colleges" (ACDE, 2019, para. 1).

The ACDE released in 2006 a set of principles regarding initial teacher education in Canada; these were updated in 2016.[5] The principles (see Figure 9.1) describe an "effective initial teacher education program" (ACDE, 2016, p. 4) and all members of the ACDE agreed to work to ensure that these principles are a part of their initial teacher education programs.

Changing School Landscape Across Canada

A significant report released in 2015 was Canada's Truth and Reconciliation Commission's (TRC) Calls to Action. The TRC's work was the result of a 7-year

---

[5] ACDE updated the Accord again in 2017. We chose to use the 2016 version for this chapter as the secondary mathematics teacher education program discussed was designed within the time of the implementation of the updated 2016 version of the Accord.

62. We call upon the federal, provincial, and territorial governments, in consultation and collaboration with Survivors,* Aboriginal peoples, and educators, to:
     i. Make age-appropriate curriculum on residential schools, Treaties, and Aboriginal peoples' historical and contemporary contributions to Canada a mandatory education requirement for Kindergarten to Grade Twelve students.
     ii. Provide the necessary funding to post-secondary institutions to educate teachers on how to integrate Indigenous knowledge and teaching methods into classrooms.
     iii. Provide the necessary funding to Aboriginal schools to utilize Indigenous knowledge and teaching methods in classrooms.
     iv. Establish senior-level positions in government at the assistant deputy minister level or higher dedicated to Aboriginal content in education.
63. We call upon the Council of Ministers of Education, Canada to maintain an annual commitment to Aboriginal education issues, including:
     i. Developing and implementing Kindergarten to Grade Twelve curriculum and learning resources on Aboriginal peoples in Canadian history, and the history and legacy of residential schools.
     ii. Sharing information and best practices on teaching curriculum related to residential schools and Aboriginal history.
     iii. Building student capacity for intercultural understanding, empathy, and mutual respect.
     iv. Identifying teacher-training needs relating to the above.

*Survivors, in this context, are those Indigenous peoples who attended residential schools.

FIGURE 9.2.   Calls to Action Related to Teacher Education (Truth and Reconciliation Commission, 2015, p. 7)

process of meeting with Indigenous peoples across Canada to document the histories of the impacts of residential schools[6] and governmental policies. The TRC outlined actions that educational systems should take in order to participate in reconciliation between Indigenous and non-Indigenous Canadians.

Across Canada, provincial Ministries of Education and Post-Secondary Education, K–12 schools, school districts, and post-secondary education programs are in the midst of addressing the Commission's Calls to Action. In particular, the two Calls to Action in Figure 9.2, from a section called "Education for Reconciliation," directly addressed the K–12 education system and teacher education programs.

Interestingly enough, two years after the work of the TRC began in 2008, the ACDE produced an Accord on Indigenous Education (ACDE, 2010), which outlined the responsibilities of Faculties of Education across the country to ensure

[6] Federally funded and church run residential schools operated in Canada for over 160 years. The last residential school closed in 1996. Approximately 150,000 First Nation, Métis, and Inuit children were removed from their families and put into the schools. The residential school system was built to solve the "Indian problem" in Canada (Royal Canadian Geographical Society, 2020).

that preservice teachers (PSTs) know about the history of residential schools and the impact that the schools and other government policies had on Indigenous peoples. The Accord outlined nine goals, listed in Figure 9.3, for Canadian teacher education programs for improving Indigenous Education.

While ACDE initiated the conversations on Indigenous education in 2010, CMEC held conversations about Indigenizing teacher education when they organized a symposium in 2018. Over the course of the symposium, the 11 recommendations outlined in Figure 9.4 were produced.

Over the last 10 years, the Canadian landscape of K–12 education and teacher education has been full of new conversations related to the relationship between Indigenous and non-Indigenous Canadians. Teacher education programs and Ministries of Education across the country have been taking up the Calls to Action. These new policies along with the ACDE accords have impacted preservice mathematics teacher education programs in Canada.

- To create **respectful and welcoming learning environments** that instill a sense of belonging for all learners, Indigenous and non-Indigenous, in all post-secondary programs of which faculties, colleges, schools, and departments of education are a part.
- To create **respectful and inclusive curricula** that allow learners to experience the Indigenous world and Indigenous knowledge in a wholistic way.
- To create **culturally responsive pedagogies** so that teacher candidates and faculty may learn about and practice Indigenous pedagogies and ways of knowing.
- To create mechanisms for **valuing and promoting Indigeneity** in education in all learning environments and in the academy.
- To create **culturally responsive assessment** practices that support socially just relations within and beyond the classroom for Indigenous and non-Indigenous peoples.
- To **affirm and revitalize Indigenous languages**, supporting Aboriginal communities and involving other faculties in the promotion, reclamation, restoration, revitalization, and teaching of Indigenous languages.
- To support the development and extension of **Aboriginal leadership in education**, substantially increasing the numbers of First Nations, Inuit, and Métis people in leadership positions and removing institutional barriers that prevent career advancement among Indigenous educators.
- To build awareness among **non-Indigenous learners**, fostering all education candidates' political commitment to Indigenous education, such that they move beyond awareness toward action within their particular sphere of influence.
- To foster **culturally respectful Indigenous research** in order to transform Aboriginal education, teacher education, continuing professional education, and graduate programs.

FIGURE 9.3. Goals for Teacher Education Programs for Improving Indigenous Education (ACDE, 2010, p. 2)

1. Establish and maintain productive relationships and meaningful partnerships for change.
2. Make Indigenous languages integral to teacher education.
3. Work with Indigenous peoples to recognize the importance of land-based education* [footnote added], prepare teachers in Indigenous education, and challenge existing systems.
4. Share and disseminate culturally appropriate resources and information. Create spaces and provide structure for sustainability.
5. Acknowledge the Report of the Royal Commission on Aboriginal Peoples, United Nations Declaration on the Rights of Indigenous Peoples, and the Truth and Reconciliation Commission's report as resources that can provide frameworks for change.
6. Recognize Indigenous languages as critical components of success for Indigenous education.
7. Respect and include Indigenous knowledge, worldviews, culture, and history in schools.
8. Value Indigenous ways of learning and teaching in the classroom.
9. Educators and decision makers must ask new questions, explore outside of their comfort zone, and open doors for Indigenous students.
10. Integrate Indigenous worldviews and perspectives into learning environments, since Indigenous education is for everyone.
11. Cultural awareness of, respectful engagement with, and collaborative partnerships with Indigenous communities are first steps to bringing Indigenous knowledge to all learning environments.

---

* Land-based education is "conversations with the land and on the land in a physical, social and spiritual sense" (Wildcat et. al, 2014, p. II).

FIGURE 9.4.   Eleven Recommendations by the Council of Ministers of Education, Canada Related to Bringing Indigenous Perspectives into Teacher Education (2019b, p. 1)

Mathematics Teacher Initial Preparation in Canada

Initial mathematics teacher education occurs within university faculties, schools, colleges, and departments responsible for implementing initial teacher education programs. In general, Canadian teacher education programs include a set of academic courses (i.e., content courses), a set of professional teacher education courses, and field experience. Because each University designs their teacher education program[7] to meet initial teacher certification standards in their province/territory, the specific number of credits and courses for the academic courses, professional courses, and field experience vary across the country.

---

[7] A teacher education program in Canada could be a 4-year undergraduate program, an after degree program where teacher education is taken after the completion of an undergraduate degree, or a combined degree program where teacher education is combined with, for example, an undergraduate degree in Arts or Science.

Some teacher education programs in Canada now also include an Indigenous education course as one of the professional teacher education courses as a way of demonstrating their commitment to the goals identified in the ACDE's (2010) Accord on Indigenous Education and to addressing the TRC's (2015) Calls to Action. The content of the Indigenous education course varies across the country but generally includes topics such as the histories of Indigenous peoples of Canada, history of residential schooling, introduction to Indigenous knowledge systems, and the roles and responsibilities of teachers.

Mathematics teacher educators, along with other subject specific teacher educators, are looking for ways to engage with the TRC's Calls to Action; the ACDE's goals from the Accord on Indigenous Education (e.g., calls for culturally responsive pedagogies and assessment practices (2010)); the principles outlined in ACDE's Accord on Initial Teacher Education (e.g., ensuring that beginning teachers have sound knowledge of subject matter, literacies, ways of knowing, and pedagogical expertise (2016, pp. 3–4)); and CMEC's recommendations to "Integrate Indigenous worldviews and perspectives into learning environments, since Indigenous education is for everyone" and "value Indigenous ways of learning and teaching in the classroom" (CMEC, 2019b, p. 1).

What might these emphases look like for a secondary mathematics teacher education program; more specifically, what ways might these emphases unfold within a mathematics education class? In this chapter, we will look at the case of one secondary mathematics teacher education program in the province of Alberta.

SECONDARY MATHEMATICS TEACHER EDUCATION AT THE UNIVERSITY OF ALBERTA

We teach in the Bachelor of Education (BED) Secondary program at the University of Alberta (UA). The BED Secondary program is comprised of a total of 120 credit units, or 4 years of post-secondary education. Of the 120 credit units, 66 credit units are from academic courses (i.e., 22 content courses taught outside the Faculty of Education), 42 credit units in professional courses (i.e., 14 courses taught in the Faculty of Education), and 12 credit units of field experience (i.e., 13 weeks, which is equivalent to 4 courses) (University of Alberta, 2020a). For the secondary mathematics education program, the academic courses include 12 courses in mathematics,[8] 1 course in Canadian history, 2 courses in English or French language or literature, and 7 optional courses. The professional courses include topics related to second language learners, educational assessment, legal obligations in relation to the teaching profession, teaching diverse learning styles, educational technology, educational psychology, Indigenous education, language and literacy across curricula, and curriculum and pedagogy.[9]

---

[8] This is equivalent to the minimum number of mathematics courses required for a Bachelor of Science in Mathematics at the University of Alberta (University of Alberta, 2020a).

[9] We describe the 4-year undergraduate program here. UA also has a 5-year combined Bachelor of Science / Bachelor of Education program and a 2-year Bachelor of Education After Degree program.

Of the professional courses, there are three courses in mathematics curriculum and pedagogy. The first course is a prerequisite to the second and third courses and is also a prerequisite to the first field experience, a 4-week Introductory Field Experience. The first course introduces PSTs to the field of mathematics education, the specifics of the Alberta secondary program of studies in secondary mathematics, theories about learning mathematics, different teaching approaches, and short-term planning (e.g., lesson planning). During the introductory field experience, it is expected that PSTs will be mentored by a secondary mathematics teacher[10] and will have the opportunity to plan to teach specific lessons.

The second and third courses in mathematics education[11] are taught concurrently and serve as prerequisites for the 9-week Advanced Field Experience.[12] These two courses prepare PSTs for long-term planning by attending to the complexities of individual learners in mathematics classrooms, developing multiple strategies for teaching mathematics, developing resources to teach mathematics to those learners, and developing assessment practices. Preservice teachers are expected to weave together all of the content of the professional education courses into developing a practice and ontological stance in relation to being a secondary school mathematics teacher. In the Advanced Field Experience, PSTs are mentored by a secondary mathematics teacher and are expected to plan, teach, and assess units of work and assume the role of a full-time classroom teacher.

In 2018, Alberta revised its standards for teacher certification to include a standard, outlined in Figure 9.5, directly related to addressing the TRC's Calls to Action. Effective September 2019, all graduates of a teacher education program in Alberta must now meet the new standards.

It is expected that university teacher education programs in Alberta will provide the experiences required for teachers to develop and begin to apply foundational knowledge of First Nations, Métis, and Inuit peoples. In 2018, faculty in each subject area were asked to explicitly address Indigenous perspectives, peoples, and worldviews in our classes. In our experiences, most PSTs have completed the Indigenous Education course[13] prior to their first course in mathematics education and come to the first course in mathematics education wondering about the ways in which the ideas and learnings in the Indigenous Education course translate into their practices as mathematics teachers. This chapter draws on our experiences of working together over three terms in the first course in mathematics education.

The professional courses and field experience requirements are the same across all programs. We have PSTs in each program in our classes.

[10] Mentor teachers are volunteers. They must have a Bachelor of Education (or equivalent), a permanent teaching certificate, three years of successful teaching practice, and support of the school administration (University of Alberta, 2020b).

[11] These courses are most often taken during the last term of the teacher education program.

[12] The Advanced Field Experience is often completed during the last term of the teacher education program.

[13] A description of the content of this course can be found in the previous section.

A Teacher Develops and Applies Foundational Knowledge About First Nations, Métis and Inuit for the Benefit of All Students

Achievement of this competency is demonstrated by indicators such as:

a. Understanding the historical, social, economic, and political implications of:
    - ○ Treaties and agreements with First Nations;
    - ○ Legislation and agreements negotiated with Métis; and
    - ○ Residential schools and their legacy;
b. Supporting student achievement by engaging in collaborative, whole school approaches to capacity building in First Nations, Métis and Inuit education;
c. Using the programs of study to provide opportunities for all students to develop a knowledge and understanding of, and respect for, the histories, cultures, languages, contributions, perspectives, experiences and contemporary contexts of First Nations, Métis and Inuit; and
d. Supporting the learning experiences of all students by using resources that accurately reflect and demonstrate the strength and diversity of First Nations, Métis and Inuit.

FIGURE 9.5. Teacher Certification Standard to Address the Calls to Action (Alberta Education, 2018, p. 6)

Preparing to Explore Indigenous Perspectives in Mathematics Education Classes

In preparing to explore the ways in which Indigenous perspectives might be included in mathematics education classes, we first reviewed literature around engaging with Indigenous perspectives in teacher education. We paid particular attention to Madden's (2015) analysis of 23 international studies, published in 2000–2012, of the ways in which teacher educators were engaging with Indigenous Education with/in university teacher education. This review included studies of preservice and graduate education. Madden described four pathways, "learning from Indigenous traditional models of teaching, pedagogy for decolonizing, Indigenous and anti-racist education, and Indigenous and place-based education" (p. 1). Most of the articles reviewed spoke about general Indigenous education within teacher education and not specifically within subject matters. The exception to this is within the context of literacy education (Strong-Wilson, 2007), history (Marker, 2011), and science (Belczewski, 2009; Higgins, 2011; Kawagly & Barnhardt, 1998; Korteweg et al., 2010). There were no specific examples in this review related to the ways in which mathematics teacher educators were *taking up Indigenizing mathematics teacher education.* Although we found these examples about general Indigenous education useful, we were also curious as to what this means within our own teaching contexts, where Indigenous perspectives is just one aspect of the course, and where it is expected that we cover all aspects of preparing to teach mathematics and teaching itself. We asked ourselves further questions.

- How might a mathematics teacher educator begin to Indigenize their mathematics education program?
- Where, or in what ways, do the Indigenous traditions of teaching or land-based learning fall within a mathematics teacher education program?
- What role does ethnomathematics play in a mathematics teacher program?
- What role does culturally responsive pedagogy play?
- What about the Indigenous teachings are so important in order to understand or develop an understanding about knowledge systems that are seen as interconnected and not separate from one another?

We decided to start by introducing Indigenous perspectives in the introductory mathematics education course. The course runs for approximately nine weeks, with 16 sessions of two hours and 20 minutes each. Each semester, there are between 25 to 40 PSTs[14] who are in the second to fifth year of their teacher education program, and they come to the class with at least 9-credit units (3 courses) of mathematics. This course provides PSTs with opportunities to develop an understanding of how mathematics is learned and how to structure learning opportunities for secondary mathematics students.

Christopher is one of the instructors of this course, and one of the co-authors of this chapter. One personal objective of his teaching is for PSTs to understand how the nature of the learner could affect the learning opportunities the PSTs create by stressing different ways that students learn, considering their various backgrounds and languages. Indigenous cultures and experiences form part of this narrative, so he selected an article that students would read, *"The 'Verbification' of Mathematics"* (Lunney Borden, 2011). This article was selected as a way to bridge between the introductory Indigenous Education course in the teacher education program and mathematics education.

Lunney Borden (2011) focuses on the structure of the Mi'Kmag language, which is the language of the Mi'Kmaw people in Eastern Canada, and how knowledge of the language and the structure of the language was used to improve the mathematics achievements of Mi'Kmaw students. According to Lunney Borden (2011), Mi'Kmag is a verb-based language, unlike many western languages that are noun-based. For instance, in geometry, instead of using nouns such as *face* and *edge*, students would describe objects using verbs such as "it can roll" or "it can sit still" to describe the type of face that an object has, and "it will go over" to describe an edge. Our mathematics education course introduced the role of language because many studies (von Glasersfeld, 1995; Vygotsky, 1978) point to the importance of language to students' learning. For instance, both Vygotsky (1978) and von Glasersfeld (1995) describe language as the medium through which students reason. Hence, the study of how Indigenous languages affect students' learning

---

[14] The majority of the PSTs are non-Indigenous. In any one year, there may be 1 in 80 PSTs in mathematics education who identify as Indigenous.

appeared to be a good place to start making sense of how Indigenous perspectives can be integrated into the teaching of mathematics.

Then, Florence, a co-author of the chapter and a member of the Métis Nation of Alberta (a recognized Indigenous Nation in Canada), was invited to participate in one class for approximately 2 hours. Florence decided that she wanted to build on the experience of the PSTs reading the Lunney Borden article and invited PSTs to bring a cultural artifact to the class. Florence did not specify what was meant by culture or by artifact. In the class, Florence first asked PSTs to introduce themselves by describing *where they are from* (as a way of connecting to place, an important concept in Indigenous perspectives) and *why they brought the artifact that they did* in a sharing circle (which is a common practice in Indigenous education). One example of an artifact was cowboy boots because it spoke to the culture of being a rancher; another example was an artifact of a rock because it spoke to the culture of being an Indigenous storyteller; a third example of an artifact was a mixing spoon because it spoke to the culture of being a baker; a fourth example was a small carving of a fishing boat because it spoke to the culture of being from a fishing village and family. By not defining *culture* and not defining *artifact* we, as instructors and PSTs, were able to begin to develop together a notion of the ways in which the term *culture* is used. This beginning experience was quite profound; as PSTs described where they were from and heard the stories of the artifacts, they began to realize their own connection to land, and the role that the *artifact* had in signifying the connection. Not only did this conversation lead to an awareness of the connection to place, the conversations called to question the meaning of *culture* and the meaning of *artifact*.

The second activity was to ask PSTs, in small groups, to examine each artifact and talk about the mathematics that was present in the artifact. The purpose of this was to help PSTs develop an awareness of the ways in which mathematical concepts and ideas are present in human cultures and artifacts. This experience led PSTs to noticing the relationship between the *mathematics* that was present in their artifacts and the mathematics program of studies they were studying in their mathematics education class.

The class finished with Florence sharing the ways in which the activities of introducing oneself from place, the stories of the artifacts, and noticing mathematics present in the artifacts related to Indigenous perspectives or worldviews. For example, one perspective is that *everything is alive* and has spirit. The sharing of the stories of the artifacts helped PSTs to understand how *non-human* objects might hold *spirit* because each of the relationships that the human being had with the artifact and the significance of that relationship pointed to the notion of the *spirit*. Another perspective is that there is really no separation between subject areas. The follow up discussion of the *mathematics* present could not be constrained to learning in the mathematics program of studies. As PSTs were identifying the mathematics *present*, they also talked about the potential use of the artifact, which led into discussions across different subject areas.

AFTER FLORENCE'S VISIT: USING INDIGENOUS PERSPECTIVES

For Christopher, Indigenous perspectives must be skillfully incorporated into teaching and learning activities to be meaningful. They must not appear to be an "add-on" or an "after-thought" because that may offend some individuals instead of motivating them. This sentiment was expressed by one student, who stated that she would feel awkward if an activity was clearly directed towards her. Therefore, Christopher asked PSTs to be sensitive and respectful of all cultures and try to include perspectives from various cultures. Furthermore, Christopher asked PSTs to make the mathematics the focus of all activities while using learners' backgrounds as a medium through which they make sense of the subject matter. With these in mind, PSTs were asked to plan and demonstrate lesson activities using what they had learned about Indigenous perspectives in relation to mathematics. We share two examples of the work that PSTs did in their first attempt to engage with Indigenous perspectives in their mathematics lessons.

Example 1: Reflections in Real Life and a Sharing Circle

One group of PSTs used the sharing circle and learners' connection to the earth in their lesson plan. The mathematics focus for the lesson was for learners to compare the graphs of transformed functions to the graph of the original function. The PSTs introduced the lesson by inviting learners to "provide examples of reflections in real life" and were specific in wanting to use examples of symmetry in nature (e.g., the symmetry in a butterfly and of a landscape reflecting off a lake). The PSTs indicated that this was a way that they were making sense of Indigenous perspectives. For their evaluation of the lesson, the PSTs provided the following, to have peers review one another's work:

- Ask students to stand in a circle, crush their papers with their graphs, and throw them in a pile on the floor.
- Ask each student to pick up one paper from the pile and say what is correct and wrong about the work.
- Students take turns sharing their thoughts on different pieces of work going around the circle.

By suggesting examples such as the symmetry of a butterfly and the reflection of a landscape in water, the PSTs showed a consciousness of people's connection to nature and the land, which is an important component of Indigenous perspectives. By asking learners to stand in a circle to share their knowledge and understanding, PSTs are pointing to the idea of a sharing circle. This notion of a sharing circle is a common practice in many Indigenous cultures and encourages one person to speak at a time. The others in the circle are to listen to the speaker. There are other aspects of the circle that are important. In this case, the PSTs have the learners all standing and this, for Indigenous peoples, means that everyone is at the *same level*;' that is, there is no *voice* in the circle that is above another *voice*. Secondly, the practice of talking about what is correct and what is not about the mathematics also offers another important teaching in Indigenous perspectives; that is, that all people are both learners and teachers.

Example 2: Using an Historical Context

Another group of PSTs incorporated an Indigenous context to meet the mathematics program of study requirement for learners to illustrate the change in slope when the x and/or y-value changes. The PSTs used the context of the decline in the population of the buffalo across North America as European settlers moved from the East to the West, from 1790 to 1889, discussing the historical and colonial aspects of the decline. They also gave the predicted current population of buffalo (350,000), and then asked a question, "Predict if the number of buffalo will be increasing or decreasing in the next few years. Why?" In this lesson, PSTs are using an historical context that very much impacted the lives of Indigenous peoples across North America. They bring it into the present-day context where there are efforts across North America to re-introduce the buffalo. The Indigenous peoples of North America have been convincing the government authorities of the importance of the buffalo for the entire ecosystem and the buffalo are being re-introduced. The context offers a space for the discussion of mathematical concepts, such as the scale to be used on the axes of the graph and the rate of change during the time period of 1790–1889. The final question of whether or not the population of buffalo would increase or decrease in the next few years offers spaces for further discussion—bringing the historical context into a present-day discussion.

What We've Learned So Far on This Journey

The two examples, presented above, represent two broad classes of examples of how PSTs incorporated Indigenous perspectives or Indigenous contexts in planning lessons. Now we examine what we have learned as we've started to implement Indigenous perspectives, or foundational knowledge of First Nations, Métis, and Inuit in our course.

The first lesson is that there is a contradiction in naming and introducing Indigenous peoples and perspectives. The contradiction is that while Indigenous peoples have been in this place called Canada for many thousands of years, the conversations and examples of Indigenous peoples are often considered from historical contexts. Focusing on the historical contexts then suggests that Indigenous peoples are no longer present. The dilemma in mathematics education, and teacher education more broadly, is to remember that the historical contexts of Indigenous peoples impact the current contexts of not only Indigenous peoples but all peoples who live in Canada. We were reminded of this when we examined the historical context example (Example 2) from the PSTs.

A second lesson we have learned is the complexities that exist in the implementation of Indigenous perspectives. The implementation of Indigenous perspectives is much more complex than the implementation of other new innovations (or teaching practices) that we've experienced in our careers as mathematics educators. For example, Florence remembers when encouraging students to talk in mathematics classes became so important or when using a different manipulative, such as algebra tiles, was introduced to teach concepts. Throughout all of the experiences of implementing some *new innovation* in a teaching practice, the question of worldview was never questioned. There were long standing shared understandings of *what school mathematics is* in the conversations. The innovations in teaching practices were always rooted in offering different opportunities for learners to better understand mathematical concepts. As mathematics educators, we could make sense of the innovations in relation to the theories that we ascribed to about learning mathematics. Indigenous perspectives now invite us to think about how is it that we think about learning mathematics. What happens when the foundational structure of our mathematics programs of study, that have been built on theories of what it means to learn mathematics, might be challenged? For example, in our experiences, one must learn whole numbers prior to learning rational numbers in our programs of study. These *learning trajectories* of content are outlined based on assumptions of what it means to learn; the trajectories are often linear, based on the perceived complexities of the content. What happens if the learning trajectory was not linear but more circular? Then, how might the programs of study be written and what might teaching practice look like?

A third lesson we have also learned is that an instructor does not need to have a wealth of knowledge (i.e., content knowledge) about Indigenous perspectives to begin to integrate such aspects into teaching and learning activities within their mathematics education classes. Christopher started assisting students in incorporating Indigenous perspectives into teaching activities while having little to no knowledge of these perspectives himself. Acting with humility, Christopher was first to admit this deficiency and called upon a more knowledgeable other, Florence, to help provide some relevant content to students. Working with a more knowledgeable other also provided an opportunity for him to learn about Indigenous perspectives, which allowed him to follow through with the PSTs as they began to engage with the content. Also, Christopher relied on his knowledge and teaching experience to guide his actions. With his limited content knowledge of Indigenous perspectives, Christopher acted as a facilitator, who created the opportunities and environment where PSTs could be innovative in their use of Indigenous perspectives. Here, Christopher's actions reflected the true meaning of a teacher acting as "a guide on the side" as PSTs took the initiative and negotiated novel ways to work with their knowledge of Indigenous perspectives in relation to planning for mathematics teaching.

Working with PSTs in this way brought some insights to the fore for the two of us. We noticed that the PSTs were very creative when given the opportunity to explore and make sense of their own emerging understandings of Indigenous perspectives in relation to mathematics education. We noticed that although the mathematics content remained the focus for most of the lessons, PSTs considered practices related to Indigenous pedagogies and that these pedagogical practices benefited all learners, not just Indigenous learners.

We shared our beginning experiences of exploring the ways that we, Florence and Christopher, are negotiating the engagement of Indigenous perspectives in one secondary mathematics education course. The course content for the classes has been focused on the content and pedagogical practices related to the Alberta secondary mathematics programs of study. Engaging with Indigenous perspectives, like we said earlier, is not just about the introduction of a new pedagogical practice in the teaching of mathematics. As Florence continues to learn from Indigenous knowledge keepers and Elders, we are continually reminded that we must continue to see ourselves as learners along with PSTs. We encourage all instructors and teachers to embrace a similar attitude, that is to continue to see themselves as learners, as they approach this evolving area of practice.

Instructors in mathematics education programs must recognize, and be willing to admit, that the way that mathematics learning is approached today has been shaped by theories that are influenced by colonial structures (for example, assumptions that all children at particular age groups will learn particular mathematics; so mathematics programs of studies are written in particular ways). The current documents and policies not only in Canada, but also internationally (e.g., the United Nations Declaration on the Rights of Indigenous Peoples, 2007), are raising awareness of the ways in which colonial structures have long dismissed Indigenous peoples, and we suggest, Indigenous perspectives. The degree to which an individual's level of awareness of the influence of colonial structures and the willingness of those individuals to engage in imagining ways in which Indigenous perspectives might influence educational structures in general, and mathematics education structures in particular, will influence the engagement of Indigenous perspectives in mathematics teacher education programs. The engagement of Indigenous perspectives across all mathematics teacher education programs like this will take time.

Dr. Edward Doolittle, a First Nation mathematician at First Nations University, and a dear friend, once said in a conversation that "Indigenous peoples have been around for thousands of years, 100 years is not a long time" when asked about the length of time it might take for the implementation of Indigenous perspectives in mathematics education. We have interpreted his statement to suggest that this work of engaging with Indigenous perspectives in mathematics education will take time, and we must be prepared to recognize that we will always be learners.

REFERENCES

Alberta Education. (2018). *Alberta teaching quality standard.* Author.

Association of Canadian Deans of Education. (ACDE). (2010). *Accord on indigenous education summary.* Author.

Association of Canadian Deans of Education. (ACDE). (2016). *Accord on initial teacher education.* Author.

Association of Canadian Deans of Education. (ACDE). (2019, September 9). *About us.* http://csse-scee.ca/acde/about-us/

Belczewski, A. (2009). Decolonizing science education and the science teacher: A white teacher's perspective. *Canadian Journal of Science, Mathematics and Technology Education, 9*(3), 191–202.

Council of Ministers of Education, Canada (CMEC). (2019a, September 9). *About us.* https://www.cmec.ca/11/About_Us.html

Council of Ministers of Education, Canada (CMEC). (2019b). *CMEC symposium on Indigenizing teacher education summary report.* Author.

Council of Ministers of Education, Canada (CMEC). (2020, May 29). *Teacher mobility.* https://www.cmec.ca/685/Teacher_Mobility.html

Government of Canada. (2020a, April 19). *Indigenous peoples and communities.* https://www.rcaanc-cirnac.gc.ca/eng/1100100013785/1529102490303

Government of Canada. (2020b, April 19). *British North America Act 1867—enactment no. 1.* https://canada.justice.gc.ca/eng/rp-pr/csj-sjc/constitution/lawreg-loireg/p1t13.html

Higgins, M. (2011). Finding points of resonance: Nunavut students' perceptions of science. *in education, 17*(3), 17–37.

Kawagly, O., & Barnhardt, R. (1998). *Education indigenous to place: Western science meets native reality.* Alaska Native Knowledge Network.

Korteweg, L., Gonzalez, I., & Guillet, J. (2010). The stories are the people and the land: Three educators respond to environmental teachings in indigenous children's literature. *Environmental Education Researcher, 16*(3), 331–350.

Lunney Borden, L. (2011). The 'verbification' of mathematics: Using the grammatical structures of Mi'kmaq to support student learning. *For the Learning of Mathematics, 31*(3), 8–13.

Madden, B. (2015). Pedagogical pathways for Indigenous education with/in teacher education. *Teaching and Teacher Education, 51*, 1–15.

Marker, M. (2011). Teaching history from an indigenous perspective: Four winding paths up the mountain. In P. Clark (Ed.), *New possibilities for the past. Shaping history education in Canada* (pp. 97–112). UBC Press.

Royal Canadian Geographical Society. (2020, April 17). *Canadian geographic Indigenous peoples atlas of Canada.* https://indigenouspeoplesatlasofcanada.ca/article/history-of-residential-schools/

Simmt, E. (2018). Curriculum in Canada: A fractal interpretation using the case of Alberta. In D. R. Thompson, M. A. Huntley, & C. Suurtamm (Eds.), *International perspectives on mathematics curriculum* (pp. 103–132). Information Age Publishing.

Strong-Wilson, T. (2007). Moving horizons: Exploring the role of stories in decolonizing the literacy education of white teachers. *International Education, 37*(1), 114–131.

The Truth and Reconciliation Commission of Canada (TRC). (2015). *Honoring the truth, reconciling the future: Summary of the final report of the truth and reconciliation commission of Canada.* Author.

United Nations. (2007). *United nations declaration on the rights of Indigenous peoples.* Author.

University of Alberta. (2020a, April 15). *University calendar.* https://www.ualberta.ca/registrar/calendar.aspx

University of Alberta. (2020b, May 29). *Volunteering to mentor a student teacher*. https://fieldexperiences.ualberta.ca/mentor-teachers

von Glasersfeld, E. (1995). *Radical constructivism: A way of knowing and learning. Studies in Mathematics Education Series: 6.* Falmer Press, Taylor & Francis Inc.

Vygotsky, L. (1978). *Mind in society.* Harvard University Press.

Wildcat, M., McDonald, M., Irlbacher-Fox, S., & Coulthard, G. (2014). Learning from the land: Indigenous based pedagogy and decolonization. *Decolonization: Indigeneity, Education & Society, 3*(3), I–XV.

CHAPTER 10

REFLECTIONS ON COMMONALITIES AND CHALLENGES IN MATHEMATICS TEACHER EDUCATION ACROSS EIGHT COUNTRIES

Mary Ann Huntley
Cornell University

Denisse R. Thompson
University of South Florida

Christine Suurtamm
University of Ottawa

In this chapter, the volume editors reflect on various commonalities and differences they noticed as they read the preceding chapters, which were written by experts in mathematics teacher education in eight countries. Additionally, they comment on some approaches to mathematics teacher preparation that they found particularly interesting. The final section of the chapter highlights issues that may influence teacher education in the foreseeable future, given the COVID-19 pandemic and unrest around racial injustice that have encompassed the globe in 2020, with cascading effects on grades K–12 education and therefore the preparation of mathematics teachers.

International Perspectives on Mathematics Teacher Education, pages 215–227.
Copyright © 2021 by Information Age Publishing

This volume includes a set of chapters on mathematics teacher education in eight countries throughout the world: Brazil, Canada, France, Malawi, New Zealand, Singapore, Sweden, and the United States. The authors of each chapter describe how programs are structured and implemented to prepare teachers to teach mathematics in pre-kindergarten through grade 12 settings. Although the focus of this book is on the initial preparation of mathematics teachers, we recognize that teacher education continues throughout a teacher's career. Thus, some chapter authors have extended their discussion to include the induction years of mathematics teaching, as well as promising approaches to the professional development of more seasoned teachers.

As described in Chapter 1, although we gave guidance to authors in writing their chapters, we did not restrict them to a certain set of questions or guidelines. As a result, the authors have described mathematics teacher education in their own way. Each chapter reflects the perspectives of the authors. A different set of authors might have written a chapter with a different focus on certain aspects of mathematics teacher education in their country.

Although each chapter is unique in its content, perspective, presentation, and focus, as we read the chapters we noticed some similarities across countries regarding the underlying philosophy on which the various mathematics teacher education programs rest. For instance, many authors wrote about their view of teachers as life-long learners, the importance of prospective teachers learning about and practicing inquiry-based methods, and teacher educators being guided by and helping prospective teachers develop a reflective practitioner stance (Schön, 1987).

Authors provided detailed descriptions of mathematics teacher education in their respective countries, outlining aspects of their programs that work well. In addition, they provided candid, rich, and thoughtful perspectives on issues they are facing, noting tensions, struggles, and dilemmas, and offering some innovative ideas for addressing them. We encourage you to read the preceding chapters in their entirety.

Instead of presenting a comprehensive comparative analysis of mathematics teacher education across the eight countries, our aim here is to discuss some common issues across the countries and how teacher educators are addressing them, as well as to highlight some ideas that were articulated by the authors that we found particularly interesting or innovative. We present these ideas around seven central themes, some of which mirror the groups of questions we posed in Chapter 1. Throughout this chapter, we use specific examples from the preceding chapters to illustrate these ideas. Bear in mind that we are not presenting an exhaustive summary; that is, examples from other countries might also illustrate the ideas discussed here.

WHAT ARE THE HISTORICAL, SOCIAL, AND POLITICAL INFLUENCES ON TEACHER EDUCATION?

The authors of the chapters all framed their discussions of mathematics teacher preparation within the historical, social, and political context of their respective country. For example, in the United States, teacher education has historically been controlled by the states rather than by the federal government; although national organizations have developed guidelines for teacher preparation, the extent to which those are mandated for programs is a state governmental decision. The authors of Chapter 7 note that Brazil has long faced social and economic inequities that have influenced teacher education programs. Since 1992 in Sweden, families have been given the freedom to choose a school instead of being required to attend the neighborhood school; this political decision has had an enormous impact, as it has led to reduced sociodemographic diversity among schools.

Although countries had varied historical, social, and political contexts in designing their preparation programs, we noticed that many of the authors mentioned a focus on preparing mathematics teachers who are able to create equitable learning environments for their students. One aspect of ensuring equity and diversity was seen in some countries' approaches to incorporating Indigenous perspectives in their mathematics teacher education programs.

For example, since 2008 Brazil has initiated a Higher Education and Indigenous Degree Program to support the education of teachers who work in elementary education at Indigenous schools, and to support Indigenous students to become teachers. The pedagogical courses vary by region in order to "respect the intercultural and territorial differences of each ethnic group" (this volume, p. 147).

A second example is New Zealand, in which the country's founding document considers the country to be bicultural, with the Māori and non-Māori people having equal standing. New Zealand also has the largest group of Pacific Islanders in the world. As a consequence, teacher educators are challenged with preparing teachers to adequately meet the learning needs of students from these diverse backgrounds and to help remedy the over-representation of Pacific and Māori students in categories of low achievement and low measures of well-being. Educational initiatives such as the Māori-medium educational system focuses on the regeneration of Māori culture and language by including Māori-medium teaching and learning experiences for students, either in Māori-medium immersion programs or in bilingual units in English-medium schools. As a result, there has not only been the development of Māori-medium initial teacher education programs, but also the inclusion, within methods courses for all teachers, of strategies that include Māori language and customs.

A third example is Canada, where teacher education programs are currently responding to the country's Truth and Reconciliation Commission's (TRC) Calls to Action that include several actions educational systems should take to participate in reconciliation between Indigenous and non-Indigenous Canadians. Mathematics teacher education programs in Canada have responded in a variety of ways, and several specific examples given in Chapter 9 demonstrate how Indigenous perspectives might be incorporated in mathematics education courses for all pre-service teachers. In addition, the Canadian authors offer advice for readers when incorporating Indigenous perspectives into their mathematics teacher education programs, reflecting on their experiences from both Indigenous and non-Indigenous perspectives.

Creating culturally responsive pedagogies extends beyond incorporating Indigenous perspectives. As one example, in the United States, which has a long history of culturally, linguistically, and socio-economically diverse students, teacher candidates are encouraged to learn how to create learning environments that draw on children's social, cultural, and linguistic strengths, and to eliminate institutional barriers to learning such as tracking and ability grouping. In Sweden, mathematics teacher educators are also addressing social, cultural, and linguistic diversity, as roughly a quarter of the students are entitled to study in a mother tongue different from Swedish due to immigration and a large influx of refugees in recent years.

WHAT IS THE CONTEXT OF SCHOOLING?

The educational environment within a country influences what quality teaching might look like. This, therefore, leads to differences in various aspects of mathematics teacher preparation across countries, such as curricula, resources, populations, class sizes, geographical context, and socio-economic differences.

In the chapter on Sweden, several examples are presented illustrating ways in which the context of schooling impacts the preparation of mathematics teachers. For example, no marks are given to students until grade 6; free school lunch is provided for all students up to grade 12; computer programming is part of the curriculum at all grades (at the request of industry); and upper secondary education students (grades 10–12) and university students receive national study support, thereby reducing the need for loans and/or extensive work responsibilities. In addition, Sweden's teacher education programs are required by law to include the student perspective in decisions about program changes.

The early education of children was noted as a priority in some countries. For example, in France, "education from the age of 3 years old has been compulsory since the 'School of Trust' law was passed in June 2019 ... [which] gives more importance to the role of these first years of education" (this volume, p. 50). As another example, although preschool is not compulsory in Sweden, it became part of the formal school system in 1998, with municipalities being responsible for ensuring that children between ages 1 and 6 have access to early childhood education. In New Zealand, before their formal schooling, nearly all children attend some form of an early childhood education program, which is partially funded by the government. The priority being placed in some countries on the education of children in their early years, together with the widening research base on the mathematical capacities of young children (e.g., Björklund et al., 2020), places heavy demands on the mathematical preparation of early childhood teachers and also the teacher educators who prepare them.

In the chapter on Brazil, we see another example of how the context of schooling impacts the preparation of mathematics teachers. Lopes and Nacarato note the increasingly complex social contexts of basic education in middle- and high schools, including large class sizes (approximately 40 students), and "marked heterogeneity and wide cultural and social diversity" (this volume, p. 155). Because of this, "teachers need to have a more encompassing and critical education in order to deal with such issues" (this volume, p. 155).

With the introduction of free primary education in 1994, Malawi has faced many challenges, such as lack of facilities and high pupil/teacher and pupil/classroom ratios (an average of 60 students per teacher in urban areas, and over 150 in some rural areas). The government has introduced numerous interventions, such as double shifts in schools to address the problem of classroom space. An additional problem in Malawi is the lack of resources available for teaching. To address this, many teacher educators in Malawi place specific emphasis on preparing teachers to be flexible and responsive to their environments. They foreshadow teachers needing to find resources on their own, such as using charcoal if markers are unavailable, and using leaves from trees to substitute for green colored paper.

WHAT IS THE STRUCTURE OF MATHEMATICS TEACHER EDUCATION PROGRAMS?

The structure of mathematics teacher education programs affects prospective teachers' experiences, and thus, the very nature of their preparation. By structure, we mean the number of institutions within a country where teachers are prepared, the level at which the program is situated, the availability of novel (or specialized) teacher education programs, whether prospective teachers have access to non-traditional programs, certification requirements, and the duration of teacher education programs.

In reading the chapters, we were struck with the large variation across countries regarding the number of institutions where teachers are prepared. For instance, in the United States, initial mathematics teacher education programs are offered at numerous colleges and universities spread across states. In other countries, specific institutions are designated as teacher education providers. For instance, in New Zealand, there are 9, 16, and 19 providers of initial teacher education at the secondary, primary school, and early childhood levels, respectively. However, in Singapore, there is just one teacher education institution, the National Institute of Education (NIE), an institute of the Nanyang Technological University (NTU). NIE faculty work closely with officials within the country's Ministry of Education to provide coherent programs to address government policy.

In many countries, prospective teachers have the option of enrolling in a teacher preparation program at the baccalaureate or at the graduate level. In other countries, such as France, teacher preparation takes place only at the post-bachelor's level. In some countries there are specialized programs for mathematics teacher education that reflect the country's specific context. As one example, consider Brazil, where there are Indigenous intercultural teaching degrees, and other teaching degrees specifically targeted for rural education.

As another example, consider Sweden, where there is high demand for mathematics (and science and technology) teachers. As a result, several different routes to becoming a mathematics teacher are being developed, including training people with degrees in mathematics (or science or engineering) who want to change careers; teacher education programs for recent immigrants; and an innovative five-year combined engineering and teacher education program at KTH Royal Institute of Technology in Stockholm, in cooperation with Stockholm University, that gives graduates "both competence to work as an engineer with a pedagogical profile and as a teacher with an engineering perspective" (this volume, p. 41).

In New Zealand, there are few non-traditional pathways into teaching. The situation is quite different in other countries, such as in the United States, where alternative routes to certification (e.g., accelerated pathways to certification, programs housed outside of colleges and universities, online teacher preparation programs, programs sponsored by charter schools) have emerged over the past two decades to address teacher shortages, as well as to address prospective teachers' challenging schedules or lack of proximity to universities. Although approximately two-thirds of certified teachers in the United States enter the field through conventional college or university programs, these alternatives offer more flexible approaches to teacher certification.

The time for teachers to earn their certification and the mechanism for doing so are different across countries. For instance, in the United States, national professional organizations have developed guidelines (i.e., standards) for programs in terms of entry requirements as well as program composition, duration, and outcomes in order for teachers to be highly qualified. There is substantial academic freedom in the design and implementation of programs across institutions based on their respective state mandates. Nevertheless, near the end of a prospective teacher's university teacher education program, teacher candidates in most states seek certification by taking a series of tests related to pedagogy and/or content; requirements for successful completion of such tests is often less than what university programs expect. Some states accept certification granted by other states, but in some cases teachers would need to complete a new series of certification tests to move from one state to another.

Likewise, although variations exist across provinces and institutions in Canada, a nationwide Association of Deans of Education ensures that certain principles are components of all initial teacher education programs. A graduate of an initial teacher education program is granted teacher certification in their respective province upon successful completion of their university teacher education program and, in general, this certification is accepted across provinces.

The duration of teacher preparation programs varies across countries. We found that it is common for programs at the bachelor's level, irrespective of preparation at the elementary or secondary level, to be four (or sometimes three) years in duration. There are some exceptions, such as Malawi, where prospective primary school teachers generally complete a program in two years, spending most of the first year in college coursework and then engaging in school practice before returning to the university for reflection on the practice and continuation of subject learning.

WHO IS INVOLVED IN MATHEMATICS TEACHER PREPARATION PROGRAMS?

The structure of mathematics teacher preparation programs within a country affects who is responsible for teacher education. It is common, but not always the case, for prospective teachers to study mathematics content courses from mathematicians, and education courses from faculty members in a college or department of education. A source of tension in some countries is the fact that mathematics educators may not be in charge of mathematics pedagogy courses. For instance, in Brazil, mathematics pedagogy courses are often offered by private institutions through distance options and taught by generalists, particularly for elementary teachers.

Mathematics teacher education programs may depend on partnerships that respect the views of multiple stakeholders in the mathematics education community. To various degrees, these partnerships might involve program connections with mathematics classroom teachers, school districts, or mathematicians. Navigating these multiple contexts can be complex, as each has its own constraints, goals, timelines, and purposes.

We noticed several examples of university-school partnerships. For instance, Brazil's Scholarship Mentorship Initiative Program (Pibid) consists of teacher-training partnerships between universities and schools. These allow prospective teachers to experience the reality of mathematics classrooms by working in elementary, middle, and high schools under the supervision of licensed schoolteachers, who also participate in education programs at the university. The undergraduates, as well as the regular schoolteachers who participate in Pibid, receive a scholarship.

Similarly, one principle of initial preparation programs in Canada, as established by the Association of Canadian Deans of Education, relates to partnerships between universities and schools. These partnerships are designed for "interweaving theory, research, and practice and providing opportunities for teacher candidates to collaborate with teachers to develop effective teaching practices" (this volume, p. 199).

In Singapore, the pre-college school mathematics curriculum undergoes revision every six years by the country's Ministry of Education. To make sure new teachers are current in their knowledge and practices, the National Institute of Education adjusts its teacher preparation curriculum accordingly.

In Malawi, prospective primary school teachers' teaching practice in schools is supervised periodically by teacher educators. However, more regularly, their practice is supervised by mentors in the school who have been trained at the teacher colleges to supervise student teachers.

In addition to university-school partnerships, in many places there are strong connections between mathematics teacher educators and mathematicians in the preparation of mathematics teachers. In Canada, annual meetings of the Canadian Mathematics Education Study Group (CMESG) bring together mathematicians and mathematics education researchers to discuss issues of mathematics education, including teacher education, and to develop a respectful rapport between the two groups. In Singapore's National Institute of Education (NIE), active research mathematicians work together and have close ties with mathematics educators to prepare school mathematics teachers. In fact, mathematicians and mathematics educators are housed in the same academic department within NIE. This priority is articulated in the first component of the NIE's Vision and Mission Statement: "To be a world-class symbiosis of mathematicians and mathematics educators" (this volume, p. 100).

In contrast, there are two scientific societies in Brazil linked to the education of mathematics teachers at the secondary level. On the one hand, there is the Brazilian Mathematical Society, which includes professors who teach courses focused on pure mathematics; on the other hand, there is the Brazilian Society of Mathematics Education, consisting of university mathematics educators. Instead of the two groups working together, there is tension between the groups as the societies have different visions for what it means to learn mathematics. Another dilemma faced by teacher educators in Brazil concerns the preparation of elementary school teachers. A large number of their courses are taught in private educational institutions by lecturers who are hired to teach, and most have no connection with educational research.

HOW ARE PROSPECTIVE MATHEMATICS TEACHERS RECRUITED?

The structure of mathematics teacher preparation programs within a country affects both who is responsible for teacher education, as well as how prospective teachers are recruited. For example, recall that in Singapore the NIE is the only teacher education institution in the country. The Ministry of Education (MOE) has sole responsibility over teacher recruitment based on the country's needs. At the undergraduate level, prospective teachers are recruited from the top one-third of each cohort of the graduating class that qualifies for tertiary education. Once recruited, they undergo a rigorous interview with the MOE, and only one of eight applicants who are interviewed are accepted. For Singapore's postgraduate degree in education, prospective teachers must also pass a rigorous interview with the MOE, and upon successful completion they usually spend half a year teaching in a government school to determine whether teaching is a career they would like to pursue. If this goes well, they then enroll as full-time students at the NIE. In either case—undergraduate or postgraduate—upon graduation the NIE awards certification and subsequently assigns teachers to schools. Schools administer a confirmation process at the end of the first year of teaching based on satisfactory performance. This entire process seems to ensure that there is a sufficient supply of teachers in Singapore schools. Because the government invests heavily in teacher education by awarding prospective teachers scholarships for their university education, this significant capital investment results in their having a 3–4 year bond of service after graduation from the NIE.

Recall that in France, all teacher preparation takes place at the post-bachelor's level. According to Grapin and Sayac (this volume): "The competitive teacher recruitment examination is the core of teacher training in France" (p. 55). More specifically, prospective teachers of elementary grades in France take an entrance exam in which they must solve problems related to the mathematics they will teach and analyze samples of student work; such exams are scored at the district level with passing score levels varying by district. Primary teachers are then hired at the district level and are not free to transfer between districts for a few years. In contrast, prospective secondary mathematics teachers take a rigorous content-based entrance exam in which they are asked to solve problems from the undergraduate mathematics curriculum and prove theorems. They also complete an oral exam based on content related to the teaching curriculum. These exams are scored at the national level, so secondary teachers can be hired anywhere in the country.

In some countries, recruitment of teachers is reported to be quite challenging, and hence, entry requirements to teacher education programs are more relaxed. This seems to be the case in Brazil, where it is reported that a teaching degree is considered to be a "second-class bachelor's degree" (this volume, p. 148). Despite the fact that a teaching degree is an opportunity for individuals from lower socioeconomic circumstances to move up socially, teaching is not attractive to many young people. Likewise, over the past decade in New Zealand, scholarships have been offered by the Ministry of Education to cover tuition fees and students have been provided with a stipend, but these measures have failed to increase enrollment of prospective secondary mathematics teachers.

The United States also reports a severe shortage of teachers, especially in secondary mathematics. The Teaching Improvements Through Mathematics Education (TIME) 2000 Program at Queens College in New York City offers a promising approach to addressing this challenge. This four-year funded scholarship program recruits mathematics teacher candidates directly from high school, and offers them full tuition scholarships and experiences throughout their program, including early field work in schools, specialized courses and seminars, and stipends to attend professional meetings.

WHAT IS THE CONTENT OF MATHEMATICS TEACHER EDUCATION PROGRAMS?

Teacher education programs generally include a set of academic courses (i.e., content courses), a set of professional teacher education courses (i.e., pedagogy courses), and a set of field experiences. How these elements are structured, and with what emphasis, varies widely from country to country, and reflects the historical traditions and culture within a country.

Developing mathematics knowledge for teaching is a goal of mathematics teacher education programs, as prospective teachers often come to a program with a procedural view of mathematics and may lack a deep understanding of concepts that they will be teaching (Conference Board of the Mathematical Sciences, 2010). Across the countries represented in this book, we learned that the content preparation of teachers varies considerably, especially at the elementary school level. For instance, in Singapore, based on their choice of academic discipline courses and pedagogy studies, some elementary teacher candidates are prepared to be elementary mathematics specialists; thus, they could have stronger mathematical preparation than some prospective secondary mathematics teachers who have chosen mathematics as the equivalent of a minor.

A different situation exists in France. Graphin and Sayac outline the dilemma of preparing elementary teachers' mathematical content knowledge in light of numerous constraints:

> How does one train teachers, who generally have a very limited scientific background, significant gaps in mathematics, and sometimes also a bad experience in the subject, to teach mathematics from nursery school (age 3) to grade 5 (age 11) in only 36 hours . . . ? Even if mathematics could be introduced in other courses (e.g., analysis of teaching practices or initiation to research), it is not reasonable to think that these training components are sufficient to enable primary school teachers to teach mathematics efficiently and support pupils who struggle with mathematics. (this volume, p. 71)

Prospective teachers in some countries have the opportunity, as well as the expectation, to conduct research near the end of their programs to consolidate their learning. For instance, one goal of teacher education in Sweden is to prepare teachers for research. Since 1993, each teacher education student in Sweden is required to complete one or two independent research projects, which frequently take the form of a small empirical study using classroom observations, surveys, or interviews with teachers or students. In this way, they are expected to "add knowledge to professional practice" (this volume, p. 23).

In Singapore, prospective teachers must demonstrate substantial knowledge of mathematics at the end of their program of study by completing a capstone course in which they research and write a report/dissertation, and make a seminar presentation on a topic in mathematics. Similarly, in France, prospective teachers are introduced to educational research and are expected to keep abreast of scientific work in the field of education. In New Zealand, prospective teachers learn to engage in research into their own practices and to make evidence-informed decisions about what is best for their students. There are numerous benefits of prospective teachers learning how to conduct research on their practice; at the same time, this generates a tension between professional-practical knowledge and the demands of writing scientific reports. This is described by Christiansen et al. (this volume) as the "academization of teacher education" (p. 23).

WHAT MECHANISMS EXIST FOR TEACHER INDUCTION AND TEACHER PROFESSIONAL DEVELOPMENT?

The retention of mathematics teachers is another issue discussed by several chapter authors. Solid induction programs that are properly resourced are critical to building the teaching profession (e.g., Britton et al., 2003), as are substantive opportunities for teachers to engage in focused professional development.

For instance, policies governing the induction of mathematics teachers to the teaching profession exist in New Zealand and Singapore. In both countries, beginning teachers are mentored by more experienced teachers, and beginning teachers receive support by having reduced teaching loads. In addition, in Singapore, beginning teachers participate in professional learning communities and are inducted into the profession by the Ministry of Education and the Academy of Singapore Teachers.

In France, the Villani-Torossian training program is a new approach to the professional development of secondary mathematics teachers, which is similar to lesson study and involves the use of mathematical laboratories. These labs are in lower and upper secondary schools, and are "essentially places for the school's teachers to discuss, share their practices, and develop cooperation" (this volume, p. 70).

The situation is quite different in Brazil, where elementary school teachers are responsible for their own professional development, which is largely provided by educational entrepreneurs. According to Lopes and Nacarato, these private companies seek to fill in the gaps in their content knowledge by producing teaching materials for use in classrooms "under a perspective that is more concentrated on techniques and skills than reflection. These companies produce *how-to* manuals as a way to mitigate these conceptual gaps in teacher education, disregarding all the discussion in research about the need for more meaningful preparation and ignoring significant production in the field" (this volume, p. 152).

CONCLUSION

The chapters in this book provide rich descriptions of many aspects of mathematics teacher education in eight countries that span five continents: South America, North America, Europe, Africa, and Asia. Given the geographical spread of the countries represented in this book, coupled with the fact that mathematics teacher education is heavily influenced by social, cultural, and linguistic traditions as well as political tendencies, the chapters provide diverse perspectives on a variety of issues surrounding mathematics teacher education. This lets us explore commonalities across the countries as well as appreciate differences in approaches to, perspectives on, and practices concerning various issues regarding mathematics teacher preparation. Looking across geographic boundaries allows us to learn about what is working well in other countries, as well as to share in the struggles.

When we conceptualized this book on mathematics teacher education from an international perspective, we were unaware of events that would unfold in 2020 that would change, in quite fundamental ways, the world in which we live. We are all presently adapting to issues facing our global society that have disrupted educational systems across the world: the COVID-19 pandemic, and unrest in response to inequities in various aspects of society.

Around the world, COVID-19 has triggered a global health, economic, and societal crisis of staggering proportions, and has resulted in the closure of many of the physical spaces where education formally takes place for students at all grade levels. As a result, many schools have turned to distance learning, or hybrid (blended) approaches to learning, where students spend part of their time in school and other parts engaged in distance learning. Consequently, this has introduced challenges and raised new questions for teacher educators regarding how to safely and equitably provide education to students (c.f., National Academy of Education, 2020).

At the same time, recent events in the United States and elsewhere have further exposed the persistent reality of racism. This has resulted in protests around the globe, highlighting longstanding inequities in many educational systems and prompting many school administrators, teachers, and teacher educators to examine how racism is embedded in education and to consider necessary changes.

These two challenges in the year 2020 have led us to examine, with new eyes, the ways in which we educate children, and thus, how we prepare teachers. Educators around the globe are re-imagining their educational systems to ensure their focus on principles of equity, fairness, and expansion of opportunities for all children. Conversations across national borders might help us to collectively identify and understand issues and challenges to ensure that teachers may use social-cultural backgrounds and experiences of their students to cultivate students' mathematical knowledge. It is important to keep in mind that the authors of the chapters in this book wrote descriptions of mathematics teacher education in their respective countries in the year 2019. The descriptions they offer are snapshots in time. Mathematics teacher education across the globe is likely to undergo substantial change as we adapt to the present circumstances. It is particularly challenging to prepare mathematics teachers for an uncertain world. Also, we wonder what professional development mathematics teacher educators might need themselves in order to adapt programs to prepare teachers for changing times. At present, there is great uncertainty as to how teacher education will unfold in response to these and other crises. We encourage teacher educators to continue conversations begun in this book as we engage in the continued improvement of K–12 education and mathematics teacher education around the world.

REFERENCES

Björklund, C., van den Heuvel-Panhuizen, M., & Kullberg, A. (2020). Research on early childhood mathematics teaching and learning. *ZDM Mathematics Education, 52*, 607–619. https://doi.org/10.1007/s11858-020-01177-3

Britton, E., Paine, L., Pimm, D., & Raizen, S. A. (Eds.). (2003). *Comprehensive teacher induction: Systems for early career learning.* Kluwer Academic Publishers.

Conference Board of the Mathematical Sciences (CBMS). (2010). *The mathematical education of teachers II.* American Mathematical Society.

National Academy of Education. (2020). *COVID-19 educational inequities roundtable series summary report.* https://naeducation.org/covid-19-educational-inequities-roundtable-series-summary-report/

Schön, D. A. (1987). *Educating the reflective practitioner: Toward a new design for teaching and learning in the professions.* Jossey-Bass Higher Education Series.

BIOGRAPHIES

EDITOR BIOGRAPHIES

Mary Ann Huntley, Ph.D., is Senior Lecturer in the Department of Mathematics at Cornell University in Ithaca, N.Y. (United States). Her research involves examining the middle- and high-school mathematics curriculum from various perspectives, including the intended, enacted, and achieved curriculum, with a particular focus on the algebra strand. Awards for her scholarly work include a National Academy of Education Spencer Postdoctoral Fellowship and an American Association of Colleges of Teacher Education Outstanding Dissertation Award. She has a background in applied mathematics, and previously served as a program officer in Instructional Materials Development at the U.S. National Science Foundation. She currently serves on the executive board of the Association of Mathematics Teachers of New York State, is the international expert on mathematics curriculum for a grades K–12 STEM education project in Armenia (funded by the European Union and administered by the World Bank), and is a co-editor of the series *Research in Mathematics Education*, published by Information Age Publishing.

International Perspectives on Mathematics Teacher Education, pages 229–238.
Copyright © 2021 by Information Age Publishing
229

Christine Suurtamm is Professor of Mathematics Education at the Faculty of Education, University of Ottawa, Canada. Her research focuses on the complexity of mathematics teachers' classroom practice, with particular interest in teachers' formative assessment practices as opportunities to attend to students' mathematical thinking. She is also the Director of the Pi Lab, a research facility funded by the Canada Foundation for Innovation. She has been the Principal Investigator on several large-scale projects in mathematics teaching and learning, was the Canadian representative on the National Council of Teachers of Mathematics (NCTM) Board of Directors, and was Co-Chair of Topic Study Groups on Assessment at the past two International Congresses for Mathematics Education (ICME-12 & ICME-13). She has won several awards for research and teaching. She is a co-editor of the series *Research in Mathematics Education*, published by Information Age Publishing.

Denisse R. Thompson, Ph.D., is Professor Emerita of Mathematics Education at the University of South Florida in the United States, having retired in 2015 after 24.5 years on the faculty. Her research interests include curriculum development and evaluation, with over thirty years of involvement with the University of Chicago School Mathematics Project. She is also interested in mathematical literacy, the use of children's literature in the teaching of mathematics, and in issues related to assessment in mathematics education. She has published in the *Journal for Research in Mathematics Education*, the *Journal of Mathematics Teacher Education*, *ZDM: The International Journal on Mathematics Education*, *Investigations in Mathematics Learning*, *Journal of Educational Measurement*, and the practice-based journals of the National Council of Teachers of Mathematics. She is a co-editor of the series *Research in Mathematics Education*, published by Information Age Publishing.

AUTHOR BIOGRAPHIES

Tariq Akmal is the Chair and Director of Teacher Education of the Department of Teaching and Learning at Washington State University, United States. Dr. Akmal has been actively involved in teacher preparation programs and licensure for the last 20 years, working closely with schools, teachers, and the state on a variety of issues. A former middle grades English/Language Arts and Social Studies teacher, he is keenly interested in the preparation of teachers. He is the past President of the Washington Association of Colleges for Teacher Education (WACTE) and the West Region Representative of the Advisory Council of State Representatives (ACSR) of the American Association of Colleges for Teacher Education (AACTE). Dr. Akmal's research examines teaching and teacher qualifications through preparation and inservice development processes, and specifically focuses on middle grades and schools. More recently, he has been collaborating with mathematics educators and looking at teacher preparation K–12. He is currently a Co-PI on two grants, a $1.1 million grant funded by the U.S. Department of Education to provide institutional support for teacher education candidates who are first generation students, from low income backgrounds, underrepresented populations, and/or have disabilities. He is also a Co-PI on a Washington State Alternative Route Block Grant that designed and implemented teacher licensure for Indigenous teacher candidates. He regularly publishes in a variety of journals, with the goal of affecting teacher development practices.

Glenda Anthony, Professor of Mathematics Education, is a co-director of the Centre for Research in Mathematics Education at Massey University, New Zealand. Working across all sectors of education, the defining thread that binds her research is the drive to understand how we can make the learning of mathematics more engaging, inclusive, and relevant. Her research work focuses on effective pedagogical practices both in the mathematics classroom and initial teacher education contexts.

Raymond Bjuland has served Mathematics Education with distinction since 1990. He is Full Professor in Mathematics Education at the University of Stavanger, Norway. He has published in a wide range of areas such as collaborative problem solving, mathematical reasoning, semiotics resources (gestures), teacher identity, mathematical knowledge for teaching, the work of teaching mathematics, lesson study, mathematical discourse, and dialogical approaches to classroom discourses. Bjuland has been the Principal Investigator in a larger cross-disciplinary research project, *Teachers as Students* (TasS, 2012–2015), supported by the Norwegian Research Council where Lesson Study was implemented in the field practice component in two Norwegian teacher education programs. Bjuland is also a research group member in two NORHED-projects: 1) (2013–2018), aimed at improving quality and capacity of mathematics teacher education in Malawi; and 2) (2017–2021), aimed at improving numeracy in Malawi primary schools.

Christopher Charles is an Assistant Professor of Mathematics Education at the University of Manitoba, Canada, with over 20 years' experience teaching mathematics at the secondary school level. Christopher completed his Ph.D. in mathematics education in March 2020 from the University of Alberta, Canada. While a graduate student, Christopher taught the first course in mathematics education at the University of Alberta six different times over three years. This is the course from which examples in the chapter are drawn.

Iben Maj Christiansen is an Associate Professor of Mathematics Education at the Department of Mathematics and Science Education, Stockholm University, Sweden. Her main research project falls within the TRACE project, which aims to trace mathematics teacher education in the practices of newly graduated teachers and investigate its recontextualization. She heads up the Swedish research school REMATH on practice-based mathematics teacher education.

Anette de Ron is a Ph.D. student of Mathematics Education at Stockholm University, Sweden. Her research interest is in mathematical problem solving and teacher education. She is involved in the TRACE project.

Andreas Ebbelind is a Senior Lecturer focusing on student teachers' professional identity development in Sweden. As a lecturer and course developer at a teacher education program, his main focus is on early childhood education.

Raewyn Eden is a Lecturer in Mathematics Education and Coordinator of the Graduate Diploma of Teaching (Primary) at Massey University, New Zealand. Drawing on her professional practice as a classroom teacher, prevalent themes in her work include how collaboration can promote teacher learning and issues of equity of educational opportunity, including how teachers can be supported to engage in ambitious teaching that promotes rich learning for diverse (all) students.

Susanne Engström is a Senior Lecturer of Technology Education at KTH, Royal Institute of Technology in Stockholm, Sweden. Her main research falls within technology and science education with a focus on cultural aspects in relation to subject content.

Mary Q. Foote is Professor Emerita of Mathematics Education from the Department of Elementary and Early Childhood Education at Queens College of the City University of New York (United States). Her research attends to equity issues in mathematics education and broadly stated examines issues in mathematics teacher education. More specifically, her interests are in cultural and community knowledge and practices, and how they might inform mathematics teaching practice. She is currently involved in research on two professional development projects: one supports teachers to teach mathematical modeling using cultural and community contexts in Grades 3–5; and the other supports teachers to develop more equitable instructional practices through action research projects that incorporate an examination of access, agency, and allyship in mathematics teaching and learning.

Susanne Frisk is a Lecturer in Mathematics Education at the Department of Pedagogical, Curricular and Professional Studies, University of Gothenburg, Sweden. Her main interest is mathematics teaching and learning in early mathematics education.

Florence Glanfield is a Professor of Mathematics Education, a member of the Métis Nation of Alberta (a recognized Indigenous Nation in Canada), and has been a mathematics educator for 37 years. Florence currently serves as the Vice-Provost (Indigenous Programming and Research) at the University of Alberta, Canada. Florence joined Christopher three times when he was teaching the course discussed in their chapter.

Nadine Grapin is a Lecturer at Paris-Est-Creteil University and a didactic researcher in mathematics at the André Revuz Laboratory of Didactics, France. She has taught mathematics at a secondary school for the past ten years and currently trains future secondary school-level mathematics teachers and primary school teachers during the second year of their Master's degree at ESPE (Graduate School for Teaching and Education) in the Creteil district academy. She also contributes to the training of full-fledged teachers. Her research focuses on the learning and teaching of whole numbers at primary school. She is particularly interested in assessment of primary school students' mathematical knowledge and has developed several studies with Nathalie Sayac on this topic, especially about the mathematical content of large-scale national assessments, the specificities of question formats (such as multiple-choice questions), and assessment support (paper-pencil compared to tablet computer). To better identify the requirements for training primary school teachers and to better understand their practices, she also studies the content of mathematical textbooks, especially for the early primary years, on the subject of teaching pupils to read and write whole numbers.

Everton Lacerda Jacinto is a Postdoctoral Researcher in Mathematics Education at the Werklund School of Education, University of Calgary, Canada. He completed his Ph.D. in Mathematics Education at the University of Stavanger, Norway. He has previously worked as a Visiting Professor in Mathematics Education at the Faculty of Arts and Education—National University of East Timor, Asia. His primary interests are education of mathematics teachers, adult education, and teaching and learning of mathematics. Thus far, his research has focused on teacher education (teaching knowledge) on the basis of the theory of Mathematical Knowledge for Teaching and on the development of pedagogical activities rooted in the historical-cultural perspective and the theory of activity. Everton Lacerda Jacinto holds a BSc in Mathematics (2007) and a MSc in Education in Science and Mathematics (2011) from the Federal University of Goias, Brazil.

Arne Jakobsen is Full Professor of Mathematics Education at the University of Stavanger, Norway. He has experience from many international research and development projects in Mathematics Education, using lesson study in professional development of teachers and teacher educators. He has successfully been co-leader of the project, *Improving Quality and Capacity of Mathematics Teacher Education in Malawi* (2014–2018), and now co-leader of the project, *Strengthening Numeracy in Early Primary Education through Professional Development of Teachers in Malawi* (2017–2021), both supported by the Norwegian Agency for Development Cooperation under the Norwegian Program for Capacity Development in Higher Education and Research for Development. He is also co-leader of the Malawi-Norway Mobility Programme in Mathematics and Mathematics Education (2019–2023), sponsored by the Norwegian Agency for International Cooperation and Quality Enhancement in Higher Education. His interests are mathematics, mathematical knowledge for teaching, lesson study, and quantitative studies in mathematics education.

Berinderjeet Kaur is a Professor of Mathematics Education at the National Institute of Education in Singapore. She holds a Ph.D. in Mathematics Education from Monash University in Australia. She has been with the Institute for the last 30 years and is a leading figure of mathematics education in Singapore. In 2010, she became the first Full Professor of Mathematics Education in Singapore. She has been involved in numerous international studies of mathematics education and was the mathematics consultant to TIMSS 2011. She was also a core member of the MEG (Mathematics Expert Group) for PISA 2015. She is passionate about the development of mathematics teachers, and in turn, the learning of mathematics by children in schools. Her accolades at the national level include the public administration medal in 2006 by the President of Singapore, the long public service with distinction medal in 2016 by the President of Singapore, and in 2015, in celebration of 50 years of Singapore's nation building, recognition as an outstanding educator by the Sikh Community in Singapore for contributions towards nation building.

Mercy Kazima is Full Professor of Mathematics Education at the University of Malawi. Her work includes teaching mathematics and mathematics education to undergraduate and postgraduate students, and supervising Ph.D. and master students' research projects. Mercy has vast experience in mathematics education research in the areas of mathematical knowledge for teaching, teaching mathematics in multilingual contexts, and mathematics teacher education. Mercy was co-leader of a large successful project, *Improving Quality and Capacity of Mathematics Teacher Education in Malawi* (2014–2018), and she is currently co-leader of another five-year project, *Strengthening Numeracy in Early Years of Primary Education Through Professional Development of Teachers* (2017–2021). Both projects are in collaboration with the University of Stavanger in Norway and are supported by the Norwegian program for capacity building in higher education (NORHED).

Cecilia Kilhamn works at the National Center for Mathematics Education (NCM) and is also a Researcher and Senior Lecturer in Mathematics Education at the University of Gothenburg, Sweden. Her research has mainly focused on early algebra, with a special interest in classroom interaction.

Veronica Jatko Kraft is a Lecturer in Mathematics Education at the Department of Mathematics and Science Education, Stockholm University, Sweden. Her main interest is in developing the secondary mathematics teacher education program.

Yvonne Liljekvist is Associate Professor in Mathematics Education at Karlstad University, Sweden. Her main research interests are mathematics teachers' professional development, and teaching design focusing on tasks and task environment.

Celi Espasandin Lopes is a Professor of Mathematics Education in the graduate program in mathematics and science teaching at "Universidade Cruzeiro do Sul" (UNICSUL/São Paulo/Brazil). She graduated with degrees in mathematics and pedagogy. She obtained both her master's and doctoral degrees in education from "Universidade Estadual de Campinas" (UNICAMP/Campinas—São Paulo/ Brazil). She has had visiting/post-doctoral positions in mathematics education at The University of Georgia (UGA) and Miami University in the USA. She is the coordinator of the CEPEME (Center for Research in Mathematics Education and Statistics) and is the lead faculty in the GEPEEM (Study Group for Research in Statistics and Mathematics Education). Her research interests are in mathematics and statistics education.

Adair Mendes Nacarato is a Professor of Mathematics Education in the graduate program in education at "Universidade São Francisco" (USF/SP, Brazil) and undergraduate course in Pedagogy. She graduated with a degree in mathematics. She obtained both her master's and doctoral degrees in Education from "Universidade Estadual de Campinas" (UNICAMP/Campinas—São Paulo/ Brazil). She is the coordinator of two research groups: Collaborative Group in Mathematics (Grucomat) and Stories of the Formation of Teachers Who Teach Mathematics (Hifopem). Her research interests are centered on pedagogical practice in mathematics, teacher education and teacher narratives.

Mathias Norqvist is a Senior Lecturer of Mathematics Education at Umeå University, Sweden. He graduated as a secondary-school teacher in science and mathematics in 1997. Since his doctoral degree in 2016 he has been teaching mathematics and mathematics education to both primary and secondary school pre-service teachers at Umeå University. His main research interest concerns students' mathematical reasoning and how task design can influence different types of reasoning.

Rimma Nyman is a Senior Lecturer in Mathematics Education at the University of Gothenburg, Sweden. Her research focus is interest and engagement in mathematics and teacher education.

Lisa Österling is a Senior Lecturer in Mathematics Education at Stockholm University, Sweden. Her research includes mathematics teacher education from sociological and critical perspectives. Lisa has contributed to the international WiFi-project on mathematical values (2012–2014), and is at present involved in the TRACE-project, researching traces of teacher education in the practices of new mathematics teachers (2016–ongoing). Her research interest is spaces where mathematics, policy, education and school contexts meet, and where images of desired teachers are challenged in terms of justice and access.

Hanna Palmér is Professor of Mathematics Education at Linnaeus University, Sweden. Her main research interests are mathematics teaching and learning in early mathematics education including preschool. Of special interest is problem solving in mathematics, digital technology in mathematics education, and entrepreneurial competences. She is also doing research on the professional identity development of mathematics teachers. She runs a research project on how problem solving and problem posing can be the starting point for early mathematics education. She is also involved in a study investigating how toddlers' numerical development can be facilitated and how the teaching of necessary aspects of numbers can be made meaningful for toddlers.

Anna Pansell is a Senior Lecturer of Mathematics Education at Stockholm University, Sweden. Her main research interest is mathematics teachers in their institutional contexts.

Astrid Pettersson is Senior Professor of Education and Mathematics Education, in Sweden. Astrid has many research interests. Her main focus is on assessment of knowledge in mathematics; she also studies how students solve problems in mathematics and their development of achievement.

Kerstin Pettersson is an Associate Professor of Mathematics Education at Stockholm University, Sweden. Her main research focus is university students' learning, in particular, their conceptions of 'Threshold Concepts.'

Inger Ridderlind is a Lecturer in Mathematics Education at Stockholm University, Sweden. Her main research interest is in assessment of knowledge in mathematics.

Amy Roth McDuffie is a Professor of Mathematics Education in the Department of Teaching and Learning at Washington State University, United States. Her research and teaching focuses on mathematics teacher education across the career continuum from preservice teacher education to accomplished practicing teachers. In particular, she centers her work on teachers' practices related to equitable pedagogies and teachers' use of curriculum as a resource for teaching and learning. In recent years, Dr. Roth McDuffie has served as a PI/coPI on three National Science Foundation Projects: *Teachers Empowered to Advance Change in Mathematics*, *Developing Principles for Mathematics Curriculum Design and Use in the Common Core Era*, and *Mathematical Modeling with Cultural and Community Contexts*. She served as the series editor for the *National Council of Teachers of Mathematics' Annual Perspectives in Mathematics Education* (2014–2016). She regularly publishes in journals focused on teacher education, mathematics teacher education, and mathematics teaching and learning, including *Journal of Teacher*

Education, Journal of Mathematics Teacher Education, Mathematics Thinking and Learning, and *Teaching Children Mathematics*.

Nathalie Sayac is a Professor of Didactics of Mathematics in Teacher Training Schools. She is a member of the LDAR (André Revuz Laboratory of Didactics, France). She has been teaching mathematics to primary school teachers (Paris-Est University, France) for 20 years. Since May 2020, she is the Dean of Normandy's Teacher Training School (Normandie Rouen University, France). She was a math teacher for 10 years in junior and high school before becoming involved in primary-level teacher training in mathematics. Her research focuses on mathematical teachers' practices and on teacher training practices in math. In recent years, she has been particularly interested in math assessments in France, both large-scale and classroom-based. She has developed a didactical approach to studying mathematical assessments and classroom assessments. This approach integrates French didactic work and various other works on assessments from both francophone and anglophone traditions. She is involved in many collaborative research projects and promotes teacher training through research and collaboration with practitioners (e.g., field trainers and teachers).

Christina Skodras is a Lecturer in Mathematics Education at the Department of Pedagogical, Curricular and Professional Studies, University of Gothenburg, Sweden. Her main interest is mathematics teaching and learning in early mathematics education.

Kicki Skog is a Senior Lecturer in Mathematics Education at the Department of Mathematics and Science Education, Stockholm University, Sweden. She researches teacher education in the TRACE project, with a focus on diversity and inclusion. She is also strongly involved in school development and related collaborations with mathematics teachers.

Lovisa Sumpter is a Reader in Mathematics Education at Stockholm University, Sweden and Professor at University of Oslo, Norway. Her main research interest is mathematical reasoning and problem solving, but she has also done research in affect and gender, thus working in affective, cognitive, and social perspectives. Her research focuses on students and teachers, from preschool to university level. She has been a board member for IGPME (International Group for the Psychology of Mathematics Education) for many years, and also served for Nordic and Swedish research networks. At the present moment, she is on the editorial board for Educational Studies in Mathematics (ESM) and International Journal for Science and Mathematics Education (IJSME).

Eng Guan Tay is an Associate Professor in the Mathematics and Mathematics Education Academic Group of the National Institute of Education at Nanyang Technological University, Singapore. Dr. Tay obtained his Ph.D. in the area of graph theory from the National University of Singapore. He has continued his research in graph theory and mathematics education and has had papers published in international scientific journals in both areas. He is co-author of the books, *Counting*, *Graph Theory: Undergraduate Mathematics*, and *Making Mathematics Practical*. Dr Tay has taught in Singapore junior colleges and also served a stint in the Ministry of Education. He was also a member of the Mathematics Senior Advisory Group for PISA 2021.

www.ingramcontent.com/pod-product-compliance
Lightning Source LLC
Chambersburg PA
CBHW060359220326
41598CB00023B/2969